Dissect Your Dinner
"A unique dining experience"

The menu:

1: Provided on your table: water, salt and sugar

2: Beverages: soft drink, milk

3: Appetizers: crackers, bread roll with butter and cheese

4: Salad: tossed salad with dressing

5: Main course: Salmon, chicken, plant-based meat patty
   White potato, sweet potato, green beans

6: Dessert: Lemon meringue pie, gelatin with whipped cream, chocolate cake with caramel sauce

©2022, 2024 by Ellen Johnston McHenry
All rights reserved. No portion of this book may be copied, distributed, or stored electronically without permission of the author.

The author gives permission to purchaser to make copies for use in a single classroom or homeschool.

Published by Ellen McHenry's Basement Workshop
Pennsylvania, USA
www.ellenjmchenry.com

ISBN 978-1-7374763-7-5

Other titles by this author:
The Elements; Ingredients of the Universe
The Chemical Elements Coloring and Activity Book
Carbon Chemistry
Botany in 8 Lessons
Protozoa: A Poseidon Adventure
Cells; An introduction to the anatomy and physiology of animal cells
The Brain
Rocks and Dirt
Discovering Motion
Mapping the Body with Art
Mapping the World with Art

All of these titles are available at most online book distributors and can be ordered wholesale through Ingram Content. Digital copies of the books can be purchased and downloaded on www.ellenjmchenry.com.

1: Before Dinner Begins

Good evening, and welcome to our restaurant! We offer a unique dining experience—not only do we serve the finest cuisine, we also assist you in dissecting your dinner all the way down to the molecular level! We just hired these waiters last week, so we might need to be patient as they learn their new job.

We will begin the dissection of your dinner even before the appetizers arrive. There are three edible things on your table already: water, salt and sugar. Let's start with your glass of water.

Yes, to dissect things down to the molecular level, we'll need some special equipment. Your ordinary scalpel and forceps won't be adequate. We'll need an amazing magnifying machine that will let us zoom in at ridiculously high levels of magnification, making things look up to one million times larger. In real life, we'd have to go to a lab that has an electron microscope worth tens of thousands of dollars. And even this machine might not even be good enough. We might have to use a machine that uses X-rays and needs super smart physicists to figure out what the pictures mean.

However, here on paper we can go cheap and just draw pictures. We can imagine that we have a super duper magnifier...

An expensive magnifier we can't afford

Hey—you haven't seen this little beauty in action yet! Please reserve your judgment for a moment. Let's use our magnifier to take a look at water. When seen with just our eyes, water doesn't appear to be made of anything. It's only when we magnify it several million times that we can actually see what it is.

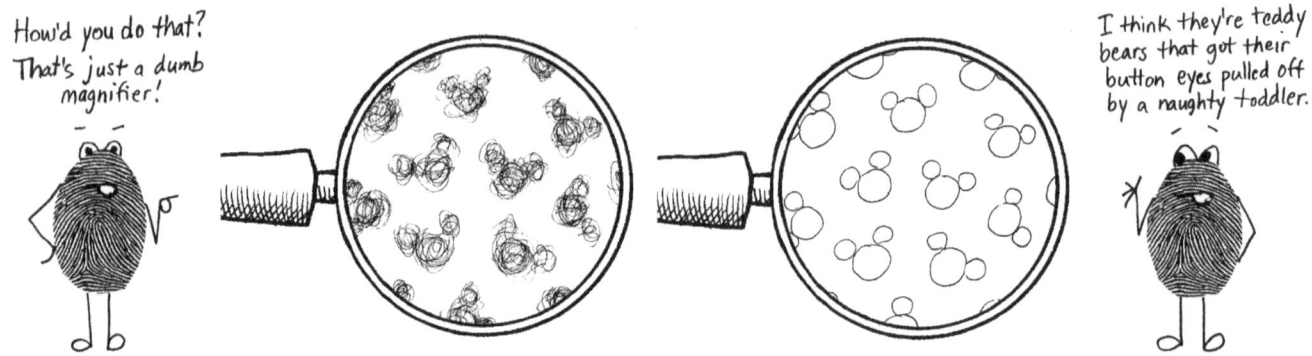

All those fuzzy blobby things are water molecules. As you can tell, water molecules are made of three parts—one large one and two little ones. The fuzzy view shows you that in reality, molecules are in constant motion so it's impossible to get them in focus. The view on the right has an artificial focus imposed upon it. (And you thought our magnifier was wimpy. It's got artificial focus!) Those Mickey Mouse shapes (yes, we knew you were thinking that) are made of three atoms: one oxygen and two hydrogens.

Atoms are the most basic particles that exist. They're a little bit like building bricks ("Legos®"). Building bricks come in many different sizes and colors and can be used to make large structures. Structures can be taken apart and the pieces can be recycled. Atoms are like the individual bricks. When we speak of a type of atom in general, we call it an *element*. Oxygen atoms can be referred to as "the element oxygen." In our building brick example, an element would be one type of brick, such as red 2x2 bricks, or white 2x6 bricks.

There are 118 different types of elements. Most of them are very useful, but some of the largest ones (numbers over 100) are very strange and only exist for a few seconds and are therefore practically useless. These 118 elements are usually written down not as a continuous list, but in a nice, neat rectangular chart called the Periodic Table of the Elements. The word "Periodic" means that there is a pattern to the way the elements are arranged, and "periodically" the pattern repeats itself. Some of these 118 elements are substances you've heard of, such as oxygen, nitrogen, hydrogen, helium, neon, carbon, calcium, magnesium, gold, silver, nickel, copper, iron and lead. Others are not so familiar and have names that look hard to pronounce. Fortunately, most of the atoms you meet in food chemistry are the easy and familiar ones, such as oxygen, hydrogen, carbon, nitrogen, sodium, and magnesium.

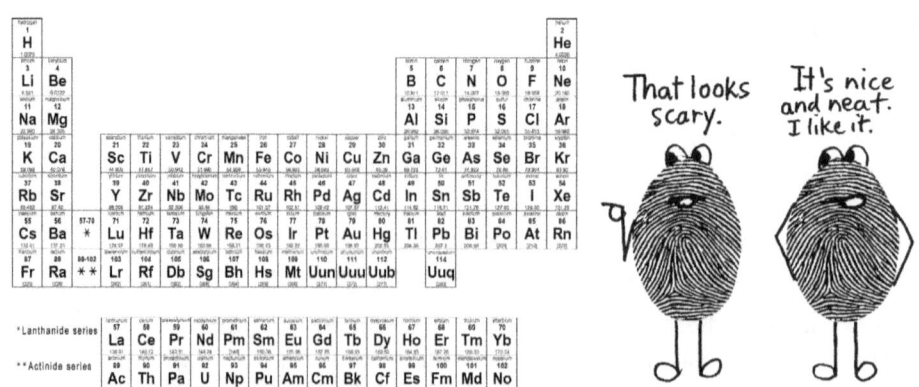

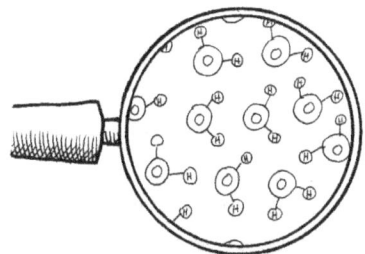

In this view, the molecules are shown as little balls stuck together with sticks. The balls are atoms. O stands for "oxygen" and H stands for "hydrogen." The sticks represent the bond that keeps the atoms together (in this case, by sharing electrons).

As you can see in our magnifier, atoms can stick together to make clumps. We call these clumps *molecules*. Here we see clumps (molecules) that are made of two hydrogen atoms and one oxygen atom. Every molecule is exactly the same. The atoms stay together because the tiny hydrogen atoms are sharing their only electron with the oxygen atom.

Just like an *atom* is a single particle of an *element*, a *molecule* is a single unit of a *compound*. *A compound is a large amount of similar molecules, with each molecule being made of at least two different elements.* Is water a compound? Yes, because all the molecules look the same, and each molecule is made of two different elements—oxygen and hydrogen. Would chicken soup be considered a compound? No, because there are so many different types of ingredients. Would pure oxygen be considered a compound? No, because even though the molecules are all the same, each molecule is made of just one element—oxygen.

So, how <u>do</u> you dissect, or tear apart, a water molecule? With a microscopic knife? Nope. Water molecules are so incredibly small that a knife would be useless. A knife blade is made of molecules that are much larger than the water molecules. It just wouldn't work. But there is a way to tear water molecules apart. We need... an electrical ZAPPER!

If we put electrodes from a battery into a glass of water and add a tiny pinch of salt or some other substance that conducts electricity, we will see bubbles forming on the electrodes. Bubbles of pure hydrogen gas will form on the negative electrode and bubbles of pure oxygen gas will form on the positive electrode. We have successfully dissected water molecules!

Does it work in reverse? If you put hydrogen and oxygen gases together would they form water molecules? Yes, they would. This is how a fuel cell works. Water molecules are split, then the gases are allowed to mix and form water again. Energy is released as the gas molecules form water molecules. The problem is that it takes energy to split the molecules in the first place, so a fuel cell can't actually create energy.

Now we're going to show you the full capabilites of our Sooper Dooper Viewer. We can zoom in using an even higher magnification and look at a single atom! Let's start with the smallest of all atoms—the hydrogen atom.

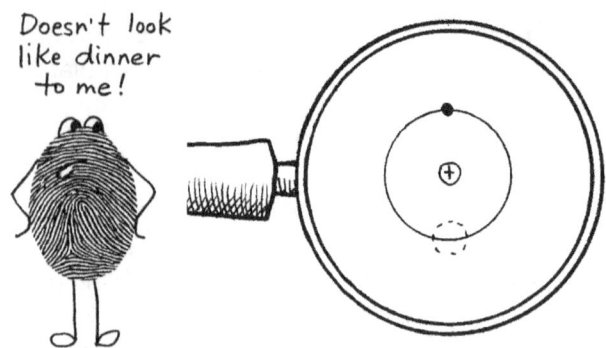

Here is a hydrogen atom. In reality, you can't actually see an atom. Atoms are just too small to see. Scientists figured out the structure of atoms using math and logic more than anything else. We draw diagrams like this one to represent atoms, but real atoms don't look like circles and dots. The dot in the center with the plus sign on it represents a **proton**. What is a proton? It's a particle with a **positive** charge (thus the plus sign). What kind of particle? That's a question for a particle physicist, not a food scientist. If we understand that a proton is a particle with a positive charge, that's enough.

The dot represents an **electron**. An electron is a particle with a **negative** charge. The circle around the proton represents an over-simplified "orbit" in which the electron travels. The electron actually whizzes around the proton in a three-dimensional way, being everywhere and nowhere all at once, looking more like a cloud than a circle. However, a circle will serve us much better as we try to understand how and why atoms stick together.

The small circle made of dashes (opposite the dot) represents an empty place that another electron could fill. Electrons love to be paired up, and hydrogen's lonely electron would love to have a partner to fill that empty space. However, if the atom takes on another electron, it will create a new problem: the atom will no longer be electrically balanced. As it is right now, the atom has one positively charged proton and one negatively charged electron. With one of each, it's balanced. If it takes on a second electron, the score will be: protons: 1, electrons: 2. The atom will have an extra negative charge, giving it an overall charge of (-1).

What should hydrogen do? It has three options it doesn't like: 1) have a lonely electron, 2) be electrically unbalanced, or 3) give its electron away to another atom. It's a no-win situation for hydrogen. Yes. such a travesty. Let's find out what hydrogen does when an oxygen atom comes along.

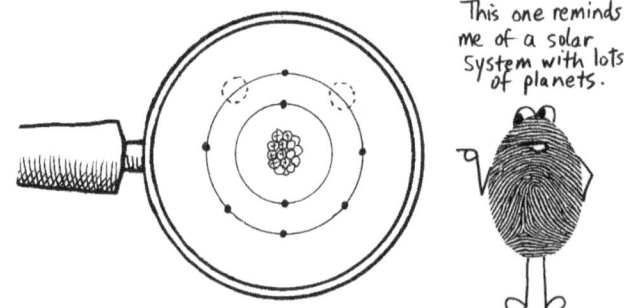

In this magnifier view, you see a representation of an oxygen atom. It's a bit more complicated than a hydrogen atom, but it's still similar. In the center, the oxygen has more than one proton; it has a whole clump of particles. There are 8 protons and also 8 **neutrons**. Neutrons are electrically **neutral**, neither positive nor negative. They just sit there. This little clump of protons and neutrons is called the **nucleus** of the atom. You'll notice that there are two rings around the nucleus, not just one. The inner ring has 2 electrons and the outer ring has 6. Those two electrons in the inner ring are very happy. They are paired up and their small ring is full with just the two of them. The outer ring is larger and can hold up to 8 electrons. (Think of it as an 8-seat minivan.) Those two dotted circles are empty "seats" that the oxygen atom would really like to fill with electrons. However, just like the hydrogen atom, the oxygen atom is faced with the problem of being electrically unbalanced if it takes on more electrons. Right now it has 8 electrons and 8 protons. If it fills those circles with extra electrons, the score will then be 10 negative electrons to 8 positive protons. What will the oxygen do?

One solution that makes both hydrogen and oxygen atoms happy is to form a water molecule. When one oxygen and two hydrogens get together they have a total of 8 electrons in their outer rings. 6+1+1 Although in this picture it looks like the oxygen has gotten all of the electrons, this is not so. The electrons can move at lightning speed (literally) and are able to circle around the hydrogens often enough to make them reasonably happy. All three atoms get the electrons circling around them just often enough to convince them that this was a pretty good solution to their problem.

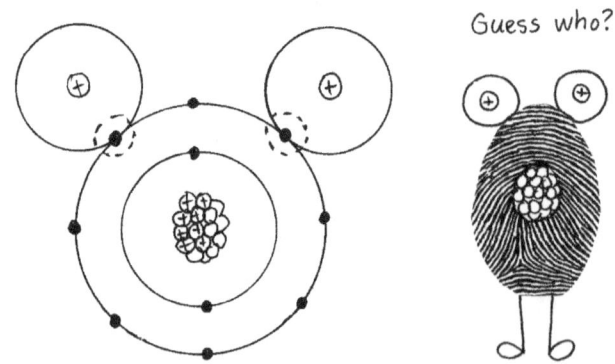

However, even though all three atoms are basically happy, this doesn't mean they are equal. The harsh reality for hydrogen is that it is puny in comparison to oxygen (or to any other atom, for that matter, since hydrogen is the smallest atom in the universe!). Hydrogen's one little proton is no match for oxygen's clump of 8 protons. Oxygen becomes a bully and begins "hogging" the electrons. This means that the electrons end up spending more time circling around the oxygen atom than they do around the hydrogen atoms. This unequal time-share of the electrons creates an imbalance in the molecule. Because the negatively charged electrons spend more time around the oxygen atom, that side of the molecule becomes slightly more negative. The side where the hydrogen atoms are stuck on becomes slightly positive because of the two protons sitting there unguarded by any electrons. Molecules like this, with a slightly negative side and a slightly positive side, are called **polar molecules**. This use of the word "polar" doesn't have anything to do with snow or bears. It simply means "having two opposite sides." The earth's poles are north and south, and function a bit like opposite ends of a magnet.

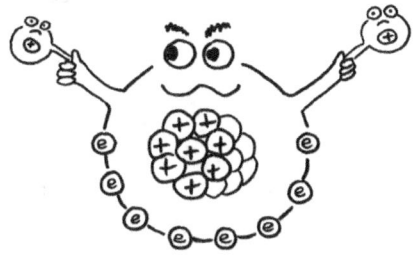

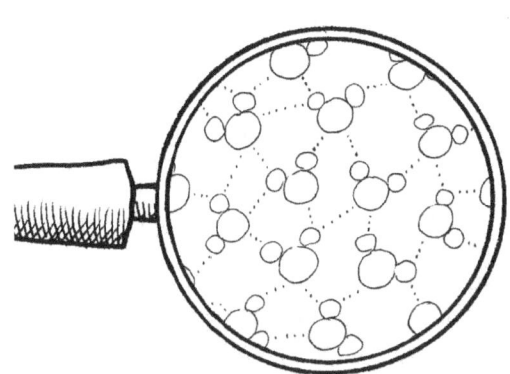

Being electrically lopsided, with a more positive side and a more negative side might seem like a bad thing, but actually it's a very good thing, and it's the reason you can take a bath or drink a glass of water. It's also the reason that plants can take up water through their roots. The negative side of one water molecule is attracted to the positive side of another water molecule. The end result is that water molecules stick together—in your bathtub, in your glass, in a raindrop, and inside plants.

This attraction between the water molecules is called **hydrogen bonding**. (At least the poor hydrogens got the bond named after them. It's compensation for getting the short end of the deal when it comes to electrons!) Hydrogen bonds are much weaker than the bonds between the oxygen and hydrogen atoms, but they are strong enough that you can see them at work. Try putting a few drops of water on a penny. Then keep adding drops until the water finally spills over onto the table. You'll be amazed at how those water molecules stick together and form a really large droplet on top of the penny! That's hydrogen bonding at work.

Some "waiters" you turned out to be! You're not very good at waiting. Be patient. The rest of this dinner won't make sense if our guests don't understand their glass of water.

There's one more very important fact about water. You'll notice that there is ice floating in your glass. Your glass of ice water demonstrates a fundamental principle of chemistry. A substance can be altered, using temperature or pressure, to turn it into a solid, a liquid, or a gas. The chemistry of the substance doesn't change, just its physical properties. Water molecules are always made of one oxygen atom attached to two hydrogen atoms, no matter whether it is ice, liquid water, or steam. When you heat water so that it turns into steam, the

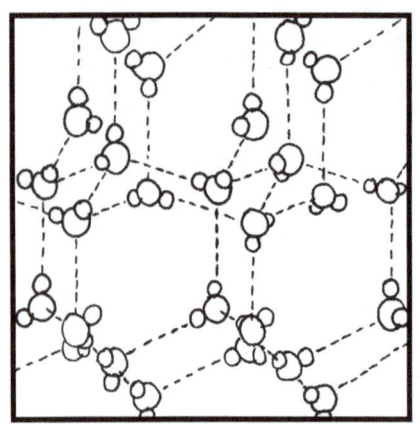

In this diagram, the dashed lines represent the electrical attraction (hydrogen bonds) holding the frozen water molecules together. These bonds (the lines) make a nice geometric pattern, and in the process they keep the water molecules farther apart than they would be if they were at room temperature and in liquid form.

water molecules themselves don't get torn apart. The heat weakens the hydrogen bonding between the molecules so that they can only form very small droplets. But the water molecules themselves remain unchanged. When water is cooled down to its freezing point, the bonds between the molecules get very strong, forming hard crystals. The water molecules actually move further apart in order to form this geometric crystal structure. When the crystals melt, the molecules move closer together again. In most other substances, it's the other way around. Usually solids are more dense (packed tightly together) than liquids. Water is backwards. It's this unusual property of water that allows ice to float instead of sink.

The scientific term for molecules being more or less tightly packed together is ***density***. Ice is less dense than liquid water because the molecules are more spread out, making fewer molecules per cubic measure. The densities of substances affect how they interact with other substances.

There's a lot more we could learn about water molecules, but we are going to move on now and look at what is in the salt shaker. If you look at salt under a magnifier, you'll see that the crystals look like little cubes. There is a reason for this, as we shall see.

Let's zoom in on the salt until we can see the molecules. Remember, this is something you can't see under a regular microscope. Our little Sooper Dooper Magnifier is much more powerful than any microscope you'd find in a biology classroom or even a medical lab.

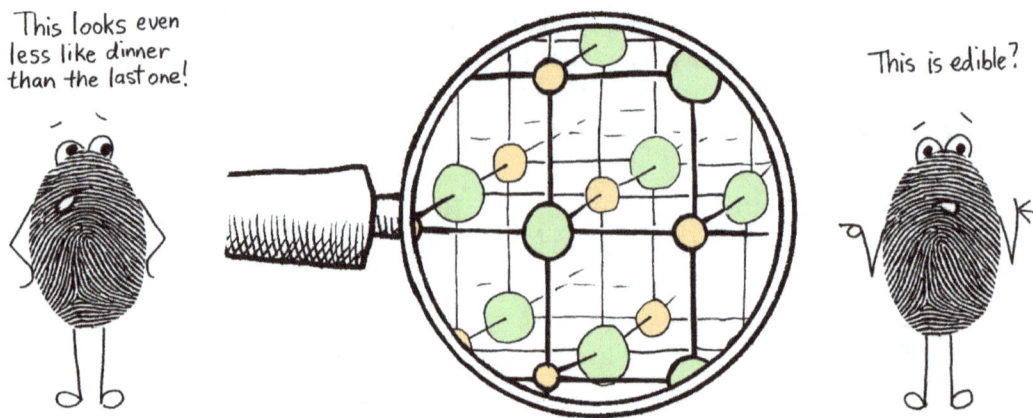

Wow—how different salt is from water! What structure! We've set our magnifier on "Ball and Stick View" so that's why you see all the circles and lines. Those circles are the atoms. The lines are the invisible bonds between the atoms (the electrical attraction). It's an endless framework of atoms all lined up in a perfectly cubic form. The atoms here aren't oxygen or hydrogen; they are **sodium** and **chlorine**. The circles that represent chlorine atoms are larger than the ones that represent sodium because chlorine atoms have more pull, or "electronegativity" than sodium atoms do, and thus they are often drawn a bit larger.

Sodium and chlorine atoms stick together because sodium has an "extra" electron it would like to get rid of, and chlorine has one empty electron space it would like to fill. Atoms don't like it when their outermost ring has either an empty spot or one lonely electron. Sodium and chlorine put their two problems together to make a solution. Sodium gives

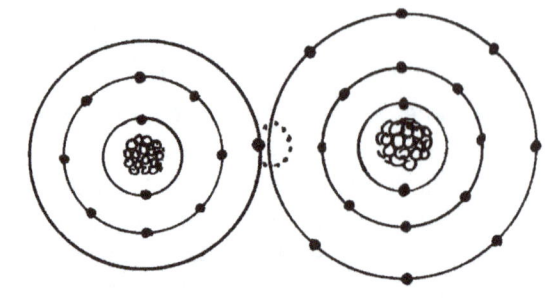

Sodium atom   Chlorine atom

its extra electron to chlorine and then they are both happy. Except that... in the process of doing this, sodium and chlorine have unbalanced themselves electrically. Sodium becomes more positive and chlorine becomes more negative. But this works out okay, because opposites attract and as long as chlorine and sodium stay next to each other everyone is (reasonably) happy.

Now... how can we dissect salt crystals? This turns out to be very easy, and you can do it without any special equipment. Just put the salt crystals into water, and presto—dissected! You won't be able to see the little atoms, though, so we'll show you an extremely zoomed-in view of dissected salt.

Remember that water molecules are "polar" and have a positively charged side and a negatively charged side. This "polarity" of water is what enables it to tear apart salt molecules. The water molecules have a stronger pull on the sodium and chlorine atoms than the sodium and chlorine do on each other. A sodium atom will leave the crystal to stick to the negative side of a water molecule. A chlorine atom will leave the crystal to stick to the positive side of a water molecule. It takes a little time for all the sodium and chlorine atoms to leave the crystal, but eventually they will all leave and the crystal will be gone. Once this has happend, we say that the salt has **dissolved** into the water. (NOTE: Sodium used to be called "natrium" so its symbol is **Na**.)

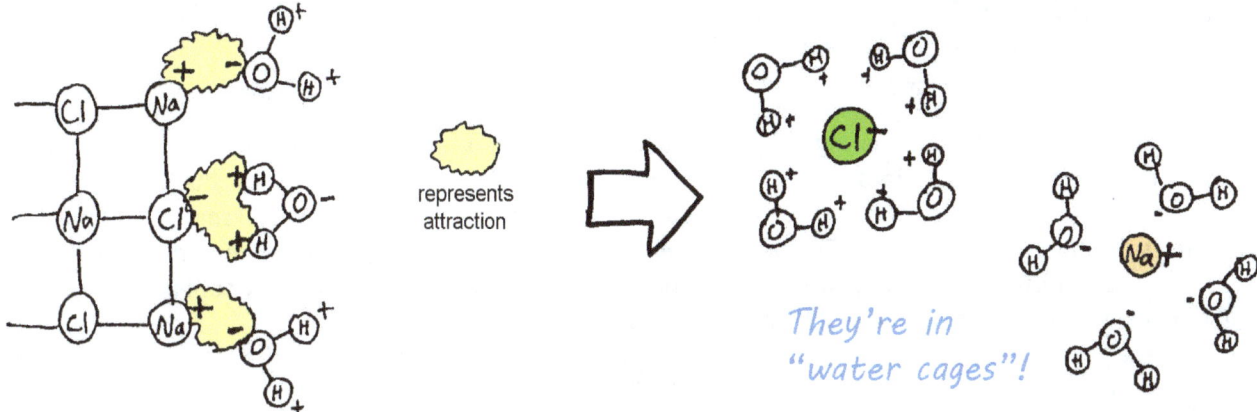

Look at the diagram on the right and notice how the water molecules surround the sodium and chlorine atoms. It almost looks like they have them imprisoned in little cages. The water molecules turn their positive sides inward to trap chlorine atoms, and they turn their negative sides inward to imprison sodium atoms. When sodium and chlorine atoms are floating around like this, unattached to anything, they are called **ions**. An **ion** is an atom that has an electrical charge. Chlorine has a negative charge because it kept that electron that it borrowed from sodium. Sodium is positive because chlorine kept its electron, leaving it with 11 protons and 10 electrons.

Atoms on the Periodic Table, are listed in their "pure" form, with an equal number of electrons and protons, before they have interacted with any other atoms. In real life, you rarely find them in this state. Atoms like sodium and chlorine are almost always found as ions, having an unequal number of electrons and protons. This can be very confusing for young chemists. You help you out, if an atom has an electrical charge, it will be written in superscript, with the plus or minus symbol to the right of the number: Na^{1+} Cl^{1-}

When an ionic substance like salt breaks down into individual atoms, or ions, we say that it has been **dissolved**. The water is called the **solvent** and the salt is called the **solute**. The salt water is called a **solution**. We'll meet some more solutions as dinner progresses.

Is the salt permanently damaged, or could the molecules be put back together again? The salt molecules can indeed be restored to their crystalline form, and this can be accomplished simply by doing nothing at all. Just let the water sit there. The water molecules will evaporate into the air and the sodium and chlorine atoms will go right back into their neat and tidy crystal lattice. (Go ahead, try it!)

So what's left to dissect before the appetizers arrive? Let's open a packet of sugar. At first glance, it might look a lot like salt—little white crystals. But if we look at them under a magnifier (just an ordinary one this time, not our Sooper Dooper one) we can see a difference right away. The salt crystals look like little cubes, but the sugar crystals don't look cubic at all. They look more hexagonal (6-sided).

salt crystals

sugar (sucrose) crystals

Now we'll switch to our amazing Sooper Dooper Magnifier and see what sugar molecules look like. But first, we'll toss the sugar crystals into our glass of water. Water has the same effect on sugar that it does on salt. (In fact, water has this effect in many substances. Water is sometimes called "the universal solvent" because of how many substances it can dissolve.) The polar water molecules pull on the sugar molecules, enticing them to leave their lovely crystal lattice and float around by themselves. So if we want so see just one sugar molecule by itself, the best way to do that is to dissolve the sugar into the water.

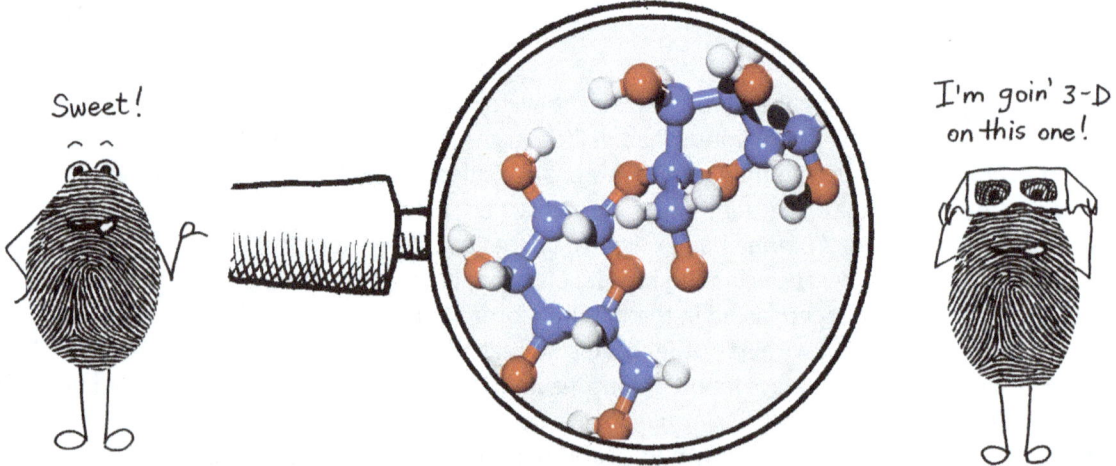

Here is just one molecule of sugar. The atoms are not lined up neatly like salt. It also looks like it should be viewed in 3D to see it properly. Some of the atoms look like they are in front or back of other atoms. To see this molecule in 3D, go to: **http://www.3dchem.com/Sucrose.asp#** Click on the molecule and it will open in a new window. If you have a touchscreen, you can interact with the molecule and rotate it. Or, you can use the commands at the bottom of this window. Look for the word "rotate," and click on the word "<u>on</u>." The molecule will begin to spin. There are many other options, too, listed at the bottom of the window. You can change from ball and stick model to other types of models. There are many ways to represent molecules. You can see a "stickless" (space-filling) model.

The picture in our magnifier shows the atoms as round balls. This kind of picture looks really nice, but you don't have a clue what those balls are, do you? For this reason, scientists have another way of representing molecules. They use letters, instead of circles or balls, to represent atoms. They keep the sticks, though. The letters they use to represent the atoms are the letter symbols found on the Periodic Table. The letters we will see most often in this book are: H for hydrogen, O for oxygen, C for carbon, N for nitrogen, Cl for chlorine, and Na for sodium.

These letter drawings don't look as artistic as the ball-and-stick ones, and they tend to look scary to non-scientists. This is the way that sugar molecule looks when drawn with letters:

The biggest "plus" about this type of drawing is that you know exactly what type of atoms you are seeing. C is for carbon, O is for oxygen, and H is for hydrogen. These three elements are the main ingredients of most of what we eat. Another big "plus" is that it's much easier to draw or print a diagram that is nothing but letters and sticks. You lose the 3D aspect of the molecule, but this downside isn't down enough, and chemistry books almost always use these letter diagrams.

Chemists get so used to seeing these types of molecules that they don't even need all the letters in their diagrams. When they see a pentagon or a hexagon, they assume that the **vertices** ("corners") are carbon atoms. Compare this diagram with the one above. Where are C's missing? Can you find a few more missing letters? There are some H's missing, also. Chemists just automatically know the C's and H's are supposed to be there.

The correct name for this molecule is **sucrose**. When we talk about putting "sugar" in a recipe, we are talking about "sucrose." In the world of science, the word "sugar" doesn't mean the stuff you bake with. "Sugar" is a more general word for a whole category of molecules that taste sweet. Sucrose is a sugar, but so are glucose, fructose, galactose, lactose, maltose, amylose and other "-ose's."

Notice that the basic structure of sucrose is a hexagon attached to a pentagon. Let's dissect sucrose by separating the hexagon and the pentagon. What is joining them? Look at the diagram and you will see that there is an oxygen atom between them. We'll have to snip off that oxygen.

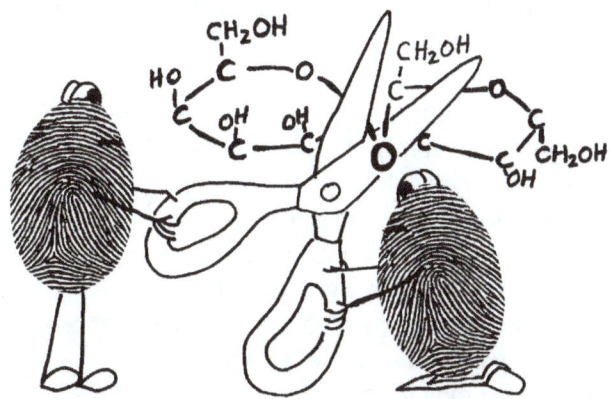

Wait a minute—SCISSORS?! Okay, okay, it works nicely in this picture and gets the point across. In real life, you need something called an **enzyme** to cut this molecule. An enzyme is a specialized protein molecule. Some enzymes act like scissors, but other enzymes act like staplers and fasten things together.

The enzyme represented here by our pair of scissors is called **sucrase**. Enzymes don't have sharp blades, of course. Enzymes are able to do their job because of their special shape.

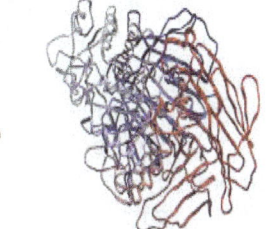

On the right is a computer-generated image of sucrase. It's a long ribbon-like molecule all twisted up into just the right shape. It doesn't look like it would be able to cut apart sucrose, does it? Yet it does, and very quickly, too.

sucrase

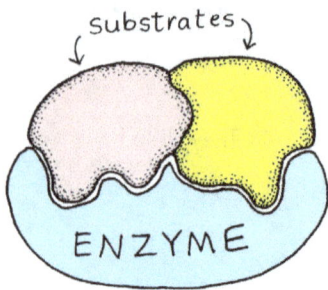

Here is the way enzymes are often look when you meet them in books. The artist makes the enzyme look like a large puzzle piece that attaches to two smaller puzzle pieces. The large piece is the enzyme and the smaller pieces are the things that the enzyme is putting together or taking apart. There's a good reason to make them look like puzzle pieces. They really do have matching shapes that fit together. The smaller pieces are called **substrates**. (Now there's a really boring science word for you. Dull, dull, dull. You'll probably forget what a substrate is by the end of this chapter.)

This is a very typical drawing of an enzyme in action. It shows an enzyme acting like a pair of scissors, cutting apart two substrates. They almost always look like oddly shaped blobs, though occasionally you'll see them as rectangles. Blobs are closer to the truth, since they actually look like a random tangle of ribbons.

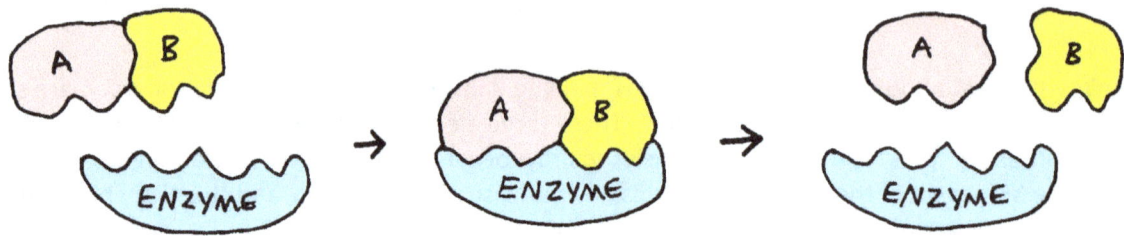

An enzyme is able to disassemble hundreds or thousands, or perhaps even millions, of substrates in its lifetime. You have sucrase enzymes in your intestines that work day and night to tear apart all the sucrose molecules you eat. Your digestive system contains many different kinds of enzymes, each one capable of tearing apart a different type of molecule.

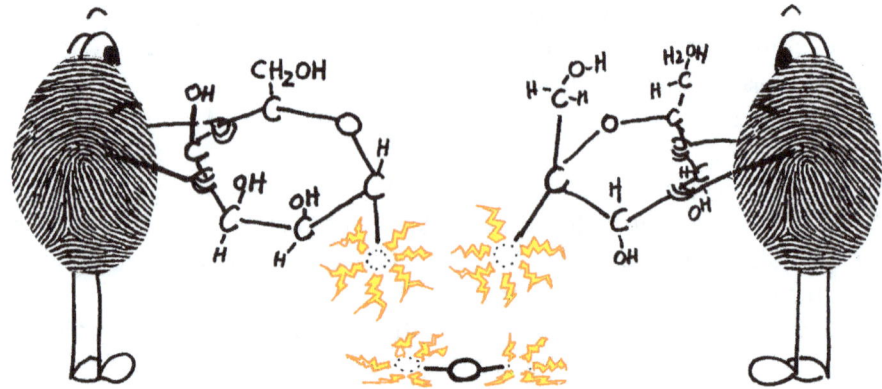

Hmmm... it looks like your new molecules aren't very happy. Neither does that snipped off oxygen down there. In fact, those broken bonds look downright dangerous with all those zappy lines coming out.

Here's what happening. Those sticks represent bonds between the atoms, right? But do you remember what a bond is? We looked at how hydrogens stick to an oxygen to make a water molecule. The "bond" was actually a place where the atoms shared an electron. It's the same with this molecule. That oxygen (O) was sharing electrons with the carbons (C) until you snipped it off. Now the oxygen is unhappy because it has two empty "holes" that are no longer filled. The carbon atoms are upset because they need to bond in four places and now they have only three of those slots filled. You've created a mess! If you walked away right now, those atoms would jump right back to where they were. If you want the molecule to stay dissected, you must patch up those broken bonds somehow. You need some spare atoms to stick onto those ragged ends. What's available?

Look! Here come some water molecules floating by. They aren't doing anything right now. Could we grab one and use it? Could hydrogens and oxygens be made into patches?

 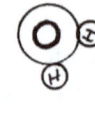

(Have you forgotten the word "substrate" yet?)

First, let's stick that snipped-off oxygen back onto one of the carbons. Now that carbon is happy again. But the oxygen is still unhappy because it is able to make two bonds and is only making one. We need an atom that only wants one bond. How about... hydrogen? Let's take a hydrogen off that water molecule and pop it onto the oxygen. There, now that molecule on the left is all patched up.

Let's try the same thing on the other side. Let's take the oxygen from water and put it onto the carbon. Then we'll patch the oxygen with the remaining hydrogen.

It looks like we've done it! We've separated the two rings and patched up all the bonds so that all the atoms are happy. What have we made? What are these rings?

We've turned sugar into... more sugar! All we've done is to turn a "two-ring" sugar into two "one-ring" sugars. These one-ring sugars are called **simple sugars** or **monosaccharides** *(mon-o-sack-a-rides)*. "Mono" means "one" and "saccharide" is a fancy word for "sugar." Sucrose is called a **disaccharide** *(di-sack-a-ride)*. "Di" means "two." (If the prefix "poly" means "many," then what would a polysaccharide look like?)

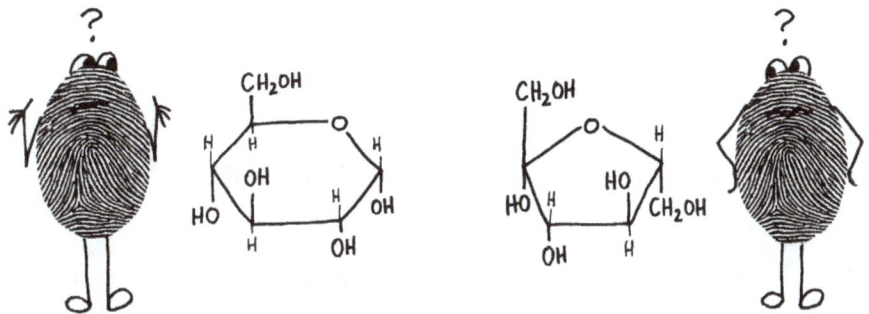

This molecule is called **glucose**. The word glucose comes from the Greek word "glukos" meaning "sugar." Not too hard. Glucose is sometimes called "blood sugar" because it's the type of sugar that floats around in your blood. It's the sugar your cells use to harvest the energy they need to stay alive. Glucose has 6 carbons, 12 hydrogens and 6 oxygens. Can you find them all? (Remember, those "corners" have invisible carbons on them!) Chemists sometimes write glucose as $C_6H_{12}O_6$, giving the numbers of each type of atom right below its symbol.

This molecule is called **fructose**. It's the kind of sugar found in fruit. Ripe fruit is sweet because it contains lots of fructose. That's easy to remember because "fruct" looks and sounds similar to "fruit." Fructose is the sweetest of all the sugars. Ounce for ounce it's sweeter than the sucrose in your sugar bowl. Fructose has the same number of each type of atom, and could also be written as $C_6H_{12}O_6$. Can you find all the atoms? Fructose is hard to find in crystal form. Most people are content to put sucrose in their dessert recipes.

Well, it looks like we've successfully dissected everything on the table. Before the beverages arrive, see if you can answer these questions. If you can, you've learned what you'll need to know to dissect the next stage of your dinner. If you can't remember, go back into the chapter and re-read until you find the answers.

1) What is an atom made of?

2) Which particles are in the center (the nucleus)?

3) What holds atoms together in a water moleclue?

4) How many bonds does oxygen want to make? (Look back at the picture where we patched up the molecules.)

5) Water is called a "polar" substance. Why?

6) What is the attraction between water molecules called?

7) Why do salt crystals come apart when you put them into water?

8) In salt water, which is the "solute," the water or the salt?

9) What do you call the things that attach to an enzyme? (Bet you forgot it already!)

10) What do enzymes do? (You are allowed to use the word you forgot in your answer to #9.)

11) What is the correct name for table sugar?

12) What is the name of the enzyme that tears apart sucrose?

13) Name the two simple sugars that link together to make sucrose.

14) What molecule can be used to patch up the ragged edges when you tear apart sucrose?

15) Did you ever notice that when you eat a lot of candy or sugar, you get thirsty? Can you think of a possible reason this might be so?

SUPPLEMENTAL VIDEOS FOR THIS CHAPTER

This curriculum has its own playlist on YouTube. Go to YouTube.com/TheBasementWorkshop and find the "Dissect Your Dinner" playlist. The videos are approximately in order, so the first ones on the list should correspond to topics from this first chapter. Come back to the playlist after you finish each chapter to watch the videos that go with those topics.

ACTIVITY 1.1   First installment of "Chew It Over," a group game to be played during a meal

This activity is designed to be something you do with family and/or friends during a meal. The questions are designed to be one of the following: informative, funny, challenging, or thought-provoking. Everyone will learn something either about science or about each other. You can use the questions in many ways. If you want to be the quiz master, you can simply read the questions out loud and see who knows the answer. Or, you could use scissors to cut them apart and then put them into a bag or bowl and go around the table letting each person draw out a question to answer. (If a question has a right answer, it is printed on the back of this page.)

CHAPTER 1: WATER, SALT, SUGAR 1) Can you name a natural substance other than water that is seen in all three states (solid, liquid, gas)?	CHAPTER 1: WATER, SALT, SUGAR 2) Which do you think uses less water, a bath or a shower?
CHAPTER 1: WATER, SALT, SUGAR 3) 90% of the world's fresh water is located on which continent?	CHAPTER 1: WATER, SALT, SUGAR 4) What % of your body weight is water? a) 1% b) 10% c) 60% d) 90%
CHAPTER 1: WATER, SALT, SUGAR 5) Plants release water vapor from their leaves. How much water does an acre of corn release in one day? (one gallon is about 4 liters) a) 4 gallons b) 40 gallons c) 400 gallons d) 4,000 gallons	CHAPTER 1: WATER, SALT, SUGAR 6) Can you guess which of these countries is NOT one of the top five producers of salt? USA, Russia, China, India, Germany, Canada
CHAPTER 1: WATER, SALT, SUGAR 7) Can you guess which one of these foods doesn't rely on salt as a key ingredient? cheese, yogurt, ketchup, mustard, soy sauce, olives, pickles	CHAPTER 1: WATER, SALT, SUGAR 8) Salt is often found underground in formations called salt domes. What other substance is usually found around or under the salt dome? a) oil b) water c) iron d) magma
CHAPTER 1: WATER, SALT, SUGAR 9) Which type of outdoor water do you like best? Ocean, lake, river, stream, puddles	CHAPTER 1: WATER, SALT, SUGAR 10) If you were required to give up either sugar or salt for one month, which would you choose?
CHAPTER 1: WATER, SALT, SUGAR 11) What is your favorite sweet food?	CHAPTER 1: WATER, SALT, SUGAR 12) What is your favorite salty food?

1) Probably not. Water is the only common substance found in all three states.
2) On average, a shower requires half as much water as a bath.   3) Antarctica
4) About 60% of your weight is water.   On average, males have 60-65%, females 50-55%.
5) 4,000 gallons of water per day!   6) Russia
7) yogurt   8) oil (meaning crude oil, or petroleum)

2: Your Beverages Arrive

The waiters have brought your beverages. They have provided milk and a carbonated drink. You may have a special name for carbonated beverages, such as "soda," or "pop" or "coke." You can imagine this to be whatever kind you like. (If you don't drink carbonated beverages in real life, just play along and pretend you do. It's just an excuse to study more chemistry.)

Let's look at the carbonated beverage first. Don't drink it yet—we need to dissect it first!

Most of a carbonated beverage is water. But there are a number of substances *dissolved* into the water. Do you remember how salt and sugar dissolved into water? The pull of the water molecules overcame the attraction that the molecules had for each other. The molecules of the **solute** (the salt or sugar) were equally dispersed among the water molecules. In this carbonated beverage we'll see that gases can also be dissolved into liquids.

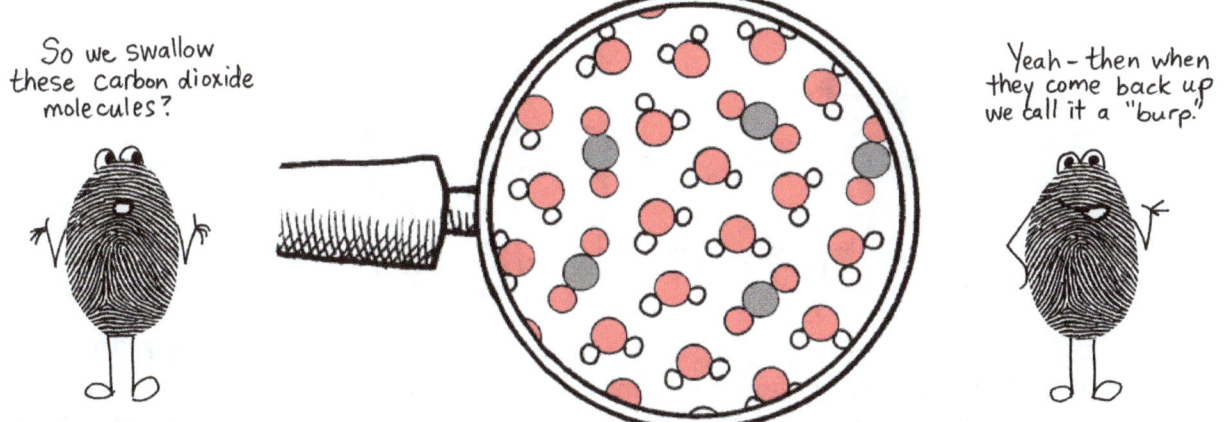

You can see the red and white water molecules, easily reconizable by their Mickey Mouse ears. (No sticks between the atoms here, just to save space.) But what are those other molecules—the ones that look like water molecules that have been straightened out? Those are **carbon dioxide** molecules, made of one carbon atom attached to two oxygen atoms. You probably know that carbon dioxide is one of the gases that you breathe out when you exhale. There is a certain amount of carbon dioxide that floats around in the air all the time. Plants take in carbon dioxide and use it for photosynthesis. You may also know that some chemical reactions, such as combustion, produce carbon dioxide. It seems strange, though, to think of carbon dioxide, a gas, being dissolved into water in the same way that salt and sugar are. Yet that is exactly what happens. (It's even weirder to think of carbon dioxide, a gas, freezing and turning into a solid. That's what "dry ice" is.)

The carbon dioxide molecule is somewhat **polar** (though you will find it in lists of nonpolar molecules because its straight line geometry makes it electrically symmetric, not lopsided). The oxygen atoms are slightly heavier than the carbon atom, and therefore they can get away with being bullies and demanding to have the electrons. The electrons end up spending more time going around the oxygens than they do the carbon. Since electrons carry a negative charge, the ends of this molecule (the oxygens) become more negative. These negative ends are attracted to the positive parts of the water molecules. The dotted lines represent this attraction, which, as you will remember, is called **hydrogen bonding**.

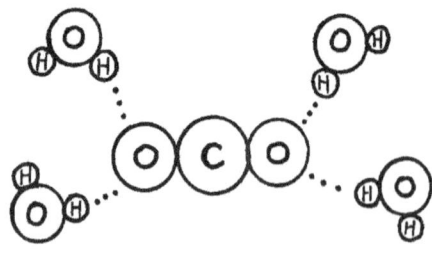

Carbon dioxide (CO_2) molecules can be dissolved into water the same way that sugar or salt can. CO_2 is the **solute** and water is the **solvent**. Here's an interesting question: can you mix and match states of matter (solid, liquid, gas) to form solutions? For example, could you dissolve a liquid into a solid? Or a liquid into a liquid? Or a solid into a gas? Oddly enough, yes, you can form a solution with just about any of these. Here are some examples of solutes dissolved into solvents.

A **gas** dissolved into a **gas**: Air, but a mixture of gases isn't called a solution. It's called a **mixture**.
A **gas** dissolved into a **liquid**: Carbonated beverages; nitrogen into blood during deep scuba dives (dangerous!)
A **gas** dissolved into a **solid**: Hydrogen can dissolve into metals, believe it or not!

A **liquid** dissolved into a **gas**: Fog
A **liquid** dissolved into a **liquid**: Vinegar (acetic acid, water); windshield washer fluid (alcohol, water)
A **liquid** dissolved into a **solid**: Gelatin; mercury dissolved into gold (the mercury seems to be solid)

A **solid** dissolved into a **gas**: Smoke
A **solid** dissolved into a **liquid**: Sugar water; salt water
A **solid** dissolved into a **solid**: Bronze (copper and tin); steel (carbon and iron)

Returning to our carbonated beverage, how is it possible to make a gas dissolve into a liquid? Well, to begin with, gases will do this on their own to some extent. For example, the water found in lakes, rivers and oceans has some oxygen. Fish and other aquatic animals "breathe" this dissolved oxygen. The way oxygen gets into the water is based on the fact that molecules are in constant motion. The bonds between the atoms in a molecule are constantly stretching and pulling and shaking, so you've got internal vibration going on. Also, the entire molecule is in motion, bumping and banging into other molecules. Molecular motion corresponds to how much "heat" a substance has. The molecules in hot substances are moving very fast. The molecules in cold objects are moving very slowly. If we cool something down to -273° C, motion stops completely. (This is called absolute zero.)

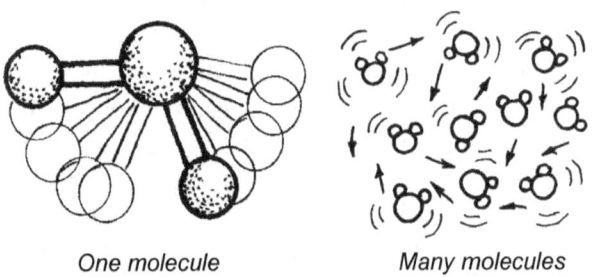

One molecule vibrating

Many molecules bumping and crashing

Oxygen molecules (O_2) are constantly going in and out of the water.

As oxygen molecules crash into the water molecules at the surface, some of them dive right in and vibrate their way down among all the vibrating water molecules. The opposite happens, too. Some water molecules move from the water to the air. Oxygen atoms in the water might go back into the air and, if conditions are right, water might move from the air back into the water. If you adjust the temperature and pressure you can control how many of each type of molecule will go in or out of the water. The faster water molecules vibrate, the more likely they are to take off and go into the air. For instance, if you turn up the heat under a pot of water on a stove, the water molecules will move faster and faster until many of them begin escaping as steam.

Regular water has dissolved gases in it, but it is certainly not fizzy like carbonated beverages. We need to pump lots and lots of gas molecules into water to get it to fizz. We must force those gas molecules to go in and stay there for as long as possible. One way to do this is to use pressure. If you squeeze the air above the water (using a machine a bit like a bicycle tire pump) you can force many more gas molecules down into the water. But this still might not be enough. You may also have to increase the amount of surface area (those places where the gas can touch water molecules) by creating lots of bubbles, like a water bubbler in a fish tank. But what if this still wasn't enough?

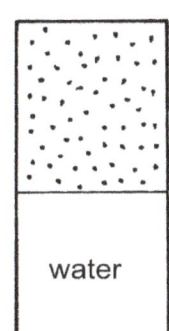
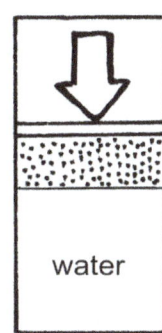

The last thing you can do to get more gas molecules into the water is to turn down the temperature. Cold water can hold more gas than hot water can. Why? Because heat is the same thing as molecular motion. The faster the molecules move, the more heat they have. The less they move, the less heat they have. So which molecules are moving faster—molecules of ice or molecules of liquid water? Liquid water, because it is warmer than the ice, and therefore its molecules are moving faster. Motion is heat. Heat is motion. More motion, more heat. Less motion, less heat.

So if we chill the water we are trying to carbonate, we will slow down the vibration of its molecules. And the slower the water molecules are going, the less they will bump into the carbon dioxide gas molecules. And the less the CO_2 bubbles are bumped, the more likely they are to stay in the water. (This is the reason that carbonated beverages go "flat" faster if they sit at room temperature than if they are kept in the refrigerator.)

What else is in a carbonated beverage, besides carbon dioxide bubbles? If you read the label, you will see that sugar, or some kind of sweetener, is a major ingredient. We'll assume that the beverage on your table has sucrose in it, which you have already dissected.

Some carbonated beverages have caffeine in them, especially colas. This is what a molecule of caffeine looks like. Caffeine is known for its ability to keep you from feeling sleepy. Food companies must think that their customers will enjoy that "wide awake" feeling and therefore want to purchase those beverages again.

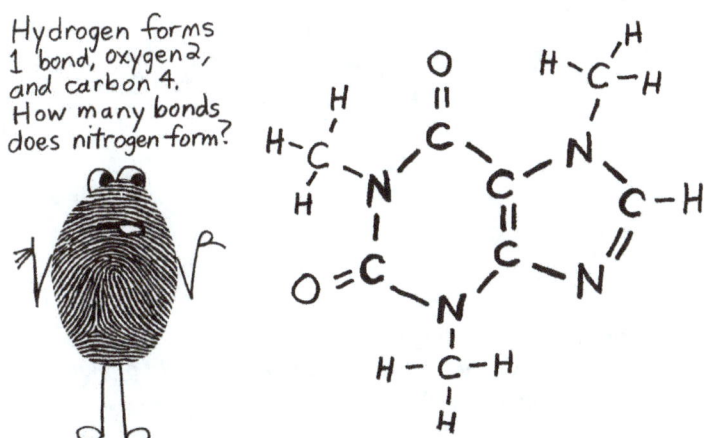

Hydrogen forms 1 bond, oxygen 2, and carbon 4. How many bonds does nitrogen form?

Caffeine has two rings: one pentagon and one hexagon. They might remind you of fructose and glucose. A big difference is that those rings include **nitrogen** atoms (N). Nitrogen molecules (as N_2) make up about 80% of the air we breathe. So what are they doing in a caffeine molecule? Just like carbon and oxygen, nitrogen is a common atom that you find in all kinds of things. Sometimes it's in a gas, and other times it's in a liquid or solid.

Caffeine is a natural substance produced by certain types of plants, especially tea and coffee plants. Plants don't need to stay awake, so why do plants make caffeine, then? For a plant, caffeine is a pesticide (a poison that paralyzes or kills certain types of insects). Fortunately, caffeine doesn't have the same effect on humans that it does on very small bugs!

What else is in carbonated beverages, besides water, fizz, sweetener and sometimes caffeine? We wouldn't want to drink them if they didn't have an enjoyable taste. Can we find any flavor molecules?

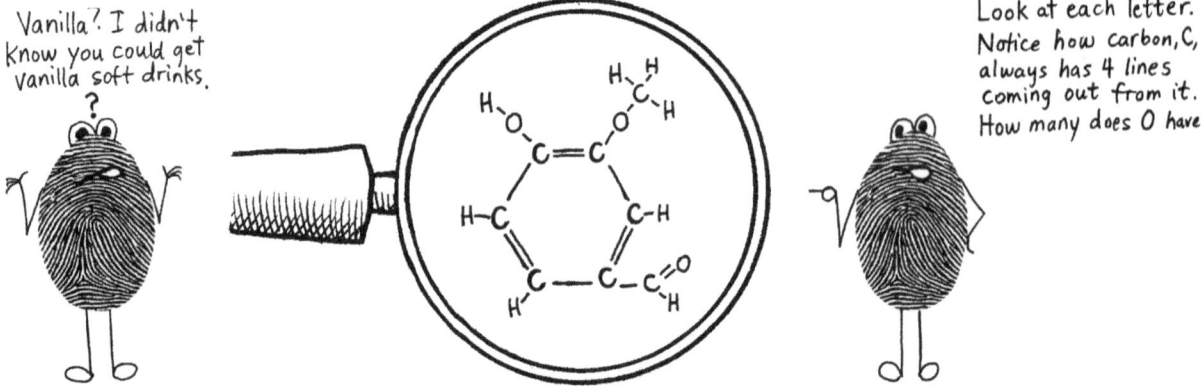

What have we got here? Looks like we've found some **vanillin**, a common artificial vanilla flavoring used in many snacks and desserts. It must be a glass of "cream soda." (That's the name for a vanilla-flavored soft drink.) Vanillin has a hexagonal ring made of 6 of carbons, with some additional carbons, hydrogens and oxygens attached to it. This hexagonal carbon ring shows up all the time in chemistry. Six carbons joined together in a hexagon shape (with a hydrogen attached to each carbon) is called a **benzene ring**. Many molecules have one or more (modified) benzene rings as part of their structure. Not all flavors have this ring; many have short strings of carbon, instead.

Benzene rings are sometimes drawn like this. Chemists know there are carbons at the corners and 6 invisible hydrogens.

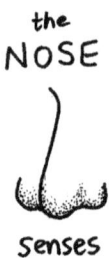

Why does this molecule taste like vanilla? Technically, it doesn't. Most of what we think of as taste is actually smell. There are only five "tastes" that the tongue can sense: sweet, sour, bitter, salty and savory. If you've tasted vanilla flavor right out of the bottle, you know that it can't be described by one of these words. Sensing flavor is a job for the nose, not the tongue. Tiny molecules of the vanillin get up into your nose and tingle receptor sites on cells inside your nose. That's why you can't taste very well when you hold your nose; taste is mostly smell. The shape of a molecule determines which receptors it can tingle. Those tingled cells send electrical impulses to the part of your brain that interprets smells. So once again, we find that **the shape of a molecule is critical to its function.**

Your carbonated beverage might also have food coloring in it. Most consumers think clear liquids are boring. They are more likely to drink something bright orange or red or yellow. Most food coloring found in carbonated beverages is artificial, meaning scientists made the molecules in a lab. This doesn't mean they are poisonous, though. All colorings used in food products have been tested thoroughly to make sure they are safe. Like any food or drink, there will always be people who have allergic reactions, or sensitivities, to them.

If you want to go natural and use color that comes from plants such as beets or carrots, the color molecules will look just as complicated.

This is "Yellow #5." Other options you can legally use are Yellow #6, Reds #3 and #40, Blues #1 and #2, and Green #3. That's it. Want orange? Mix yellow with one of the reds. Purple? Mix a blue and a red. Black? Use a lot of blue with some added red, yellow and green.

Some beverages also have **preservatives** that discourage bacteria, molds and yeasts from living in the drink. A common preservative is **potassium benzoate**. **Potassium** (symbol "K") is another type of element, and can be found on the Periodic Table at number 19. The most significant fact to know about potassium is that it has only one place it can bond, just like hydrogen. In this molecule, the potassium is hanging out with one of the oxygens.

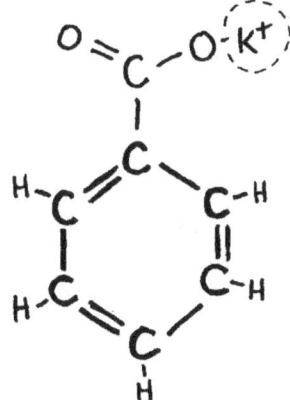

The "benzo" part of the name comes from the hexagonal ring, which can be turned into a benzene ring if the chemistry of the solution is just right. A benzene ring by itself, with no extra atoms stuck to it is a fairly dangerous molecule named **benzene**. Benzene is suspected to be a **carcinogen**—a substance that causes cancer. As long as the ring in this molecule keeps those extra atoms attached to it (a carbon, two oxygens and a potassium) it's considered to be harmless to humans. If you're a microorganism, however, you are in big trouble. When put into a liquid such as fruit juice or soda, potassium benzoate drops its potassium ion and picks up a hydrogen instead, turning itself into **benzoic acid.** The benzoic acid goes into the cells of the microorganisms and prevents them from being able to digest sugar. Basically, the little critters starve to death while being surrounded by sugar!

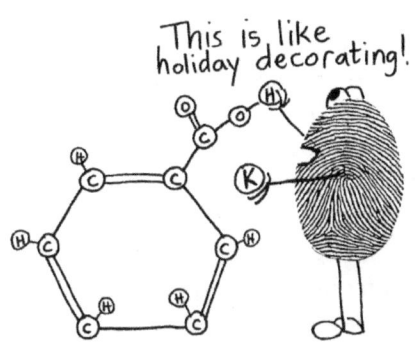

Benzoic acid doesn't affect humans the way it affects microorganisms, but there is a small potential danger to humans. If benzoic acid comes into contact with vitamin C, the vitamin C molecule strips off those extra atoms at the top (C, O and O), and thus turns benzoic acid into a benzene ring. However, the total amount of benzene formed inside a beverage can is very, very small. You get a much bigger dose of benzene by breathing the air in a big city, or by smelling gasoline fumes as you are pumping gas into your car. You'd have to drink five gallons of a carbonated beverage every day to get even close to the amount of benzene you get from other sources.

The US Food and Drug Administration runs tests on beverage products to determine if they contain unacceptable levels of benzene. Companies that produce these products are warned that they must find a way to reduce the levels down to what the FDA considers safe. In 2008, the Coca-cola® company announced that they were going to stop using benzoates in all their soft drinks except Fanta®, Dr. Pepper® and Coca-cola Zero®.

The use of preservatives is controversial, but seems to be a "necessary evil." We prefer not to have harmful substances in our drinks, but we don't want to find bacteria or mold growing in them, either. Modern food delivery systems usually require that food be able to sit in storage for a certain amount of time. Food companies get into much bigger trouble if microorganisms are found in their products than they do if they use preservatives. Of course, some products can be preserved just by keeping them cold, but this requires a lot of energy. For some products, the added energy costs would make them too expensive for the consumer.

When we talked about potassium benzoate turning into benzoic acid, we used the word "acid" without explaining it. We could do this because you are probably already familiar with the word *acid*. You know that lemon juice and unripe apples are acidic. The acid in these fruits is what gives them their sour taste. But did you know that many carbonated beverages are just as acidic as lemons, even if they taste sweet? We already mentioned benzoic acid, but you meet other acids in carbonated beverages, too. The most common one is **phosphoric acid**. Its name comes from the element **phosphorus** (P), number 15 on the Periodic Table.

Before we launch into a chemistry lesson about acids, let's ponder this question: Why do food companies put acids into carbonated beverages, anyway? They are supposed to taste sweet, not sour. Actually, there is a bit of sour "tang" to them, even if you don't notice it. Consumers prefer drinks that are sweet yet still acidic enough to make their salivary glands tingle. The acid in these beverages isn't enough to bother people who don't particularly like sour tastes, because there is so much added sugar that it covers the tartness. There is also a chemical reason to add acid to carbonated beverages: preservatives like potassium benzoate work more efficiently in an acidic environment. If you're adding potassium benzoate to natural fruit juice, the fruit provides the acid. But artificially flavored beverages need to have acid added to them in order for the preservatives to work.

Phosphoric acid in carbonated beverages is just as controversial as potassium benzoate (or its "sister" molecule, sodium benzoate). Some people love to quote the fact that phosphoric acid can be used to remove rust from metal. One Internet rumor says that cola drinks will dissolve a nail in 4 days. That'll scare you from ever drinking a cola again, eh? (It turns out to be a false claim, of course. If you want to see the results of an experiment where someone actually tried this, you can go to: **http://joshmadison.com/2003/12/14/will-coke-dissolve-a-nail-experiment/**) Pure phosphoric acid in large amounts might be able to dissolve rust or soften a nail, but the amount that is in carbonated beverages is so low that these Internet claims are ridiculous.

Phosphoric acid has also been accused of causing children to have weak bones and cavities in their teeth. They say that the phosphorus atoms are able to grab calcium atoms out of the digestive system and blood, so that they don't get delivered to the bones and teeth that need them. It is true that phosphorus atoms can grab calcium atoms, but to what degree? Enough to harm you? Studies have been done to test whether consumption of carbonated beverages (especially colas) affects bone health. Some studies claim to have found a definite link between cola consumption and reduced bone density. Other studies claim there is no link at all. How are we to know which study is right?

All researchers agree, however, that people should not drink carbonated beverages all the time. The high sugar content provides plenty of food for the bacteria that live in our mouths, and the acids in the drinks make the environment of the mouth just right for them to multiply. Even natural fruit juices can be a problem if you sip on them all day. To get rid of this extra sugar and acid, brush your teeth as often as you can.

Now it's time to find out exactly what an acid is. To do this, we'll start by looking at water again.

You'll remember that water molecules are made of two hydrogen atoms attached to an oxygen atom. Well, it turns out that those hydrogen atoms are not very faithful to their water molecules. They sometimes go wandering off, leaving H_2O as OH^-.

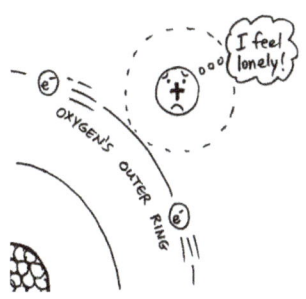

It is important to notice that once a hydrogen atom's single electron goes off to join the electrons in the oxygen atom, the hydrogen atom is reduced to being nothing but a proton. Just one proton! Can that proton still be called a hydrogen atom? Not really. We'll need to start calling it a **hydrogen ion**. An ion is atom that has become electrically unbalanced. In this case, the imbalance is plain to see, as there is just one proton with a positive charge, and no electron to balance it. Since a hydrogen ion is nothing but a proton, we can use either word and they mean the same thing. "HYDROGEN ION" = "PROTON" (This is a very useful thing to know when studying chemistry. Many students do not realize that these words are interchangeable.)

And so it happens that once in a while the lonely hydrogen ion will leave its water molecule and go off to seek its fortune elsewhere. Before long, it runs into a sad water molecule that is limping along with only one "ear." Off to the rescue it goes, and sticks itself to this disadvantaged water molecule, restoring it to H_2O. Hmm... wonder why that water molecule was missing a hydrogen? Could it be because one of *its* hydrogens got unhappy and left? Yes, hydrogens are that stupid. They keep leaving their old water molecules to join new ones even though their new molecules are identical to their old ones. The hydrogens apparently don't understand the concept that the grass really isn't greener on the other side of the fence.

  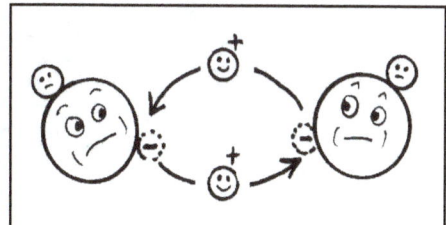

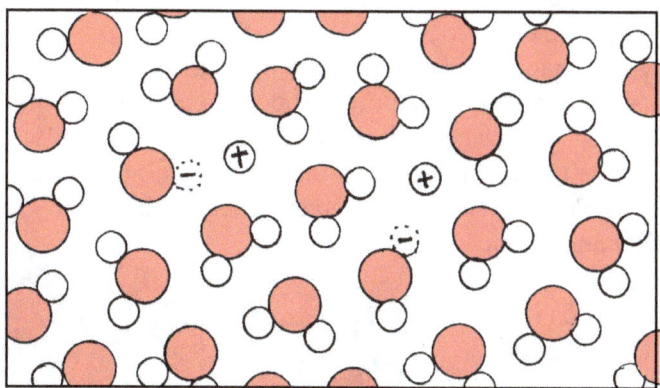

Imagine those hydrogen ions zooming around, pointlessly trading places with each other, when—SNAP! We take a picture. We have a split second of time frozen in a "photograph." Most of the water molecules are intact. But we've caught a few hydrogens mid-switch.

As you can see in this diagram, the hydrogen atom's electron stays with the oxygen atom. That little circle with the minus sign in it represents the electron that was left behind. These "broken" water molecules are no longer H_2O. They are now called **hydroxide ions**, and are written like this: OH^-. So in this picture we have lots of regular water molecules, H_2O, two hydrogen ions, H^+, and two hydroxide ions, OH^-.

In normal water, the number of hydrogen ions, H^+, always equals the number of hydroxide ions, OH^-. The ions are leaving molecules and joining molecules at about the same rate. So overall, water is electrically balanced. The positive and negative charges sort of cancel out.

What would happen if we added extra protons to regular water? It would certainly upset the balance of positive and negative ions. Is it possible to add extra protons? What about adding extra hydroxide ions?

There are certain substances that release H^+ ions (protons) when they are dissolved in water. For example, if you put hydrochloric acid, HCl, into water, the ions H^+ and Cl^- would separate. The Cl^-s would quickly be surrounded by water molecules, as we saw in the salt water. The H^+s would be your source of extra protons.

A substance that can release OH^- ions is sodium hydroxide, NaOH. ("Na" is the symbol for sodium. It used to be called "natrium.") When NaOH is put into water, you get Na^+s surrounded by water cages and loose OH^- ions all over the place.

An increase in either the number of hydrogen (H^+) ions or hydroxide (OH^-) ions in a solution affects the chemistry of the solution, so this imbalance is important for chemists to measure. They use a scale called the **pH scale**. The letters **pH** are most often interpreted as being an abbreviation for "**p**otential **H**ydrogen" because if a hydrogen ion (a proton) gets just one electron, it becomes an actual hydrogen atom again. That's why the H is capitalized; "H" is the symbol for the element hydrogen.

The pH scale runs from 0 to 14. The middle of the scale, 7, is defined as **neutral**. Numbers below 7 are **acids**. The lower the number, the more acidic the solution is. Substances that have a pH value greater than 7 are called **bases**. And, just to confuse you, bases have an alternate name, too: **alkaline** substances. You'll find these words used interchangeably in chemistry texts. One minute they'll be talking about bases and the next minute they'll be talking about how alkaline something is. These terms mean the same thing. Alkaline substances release hydroxide ions (OH^-), the counterparts (or "opposites") to the hydrogen ions (H^+).

Here are the pH values of some common household substances.

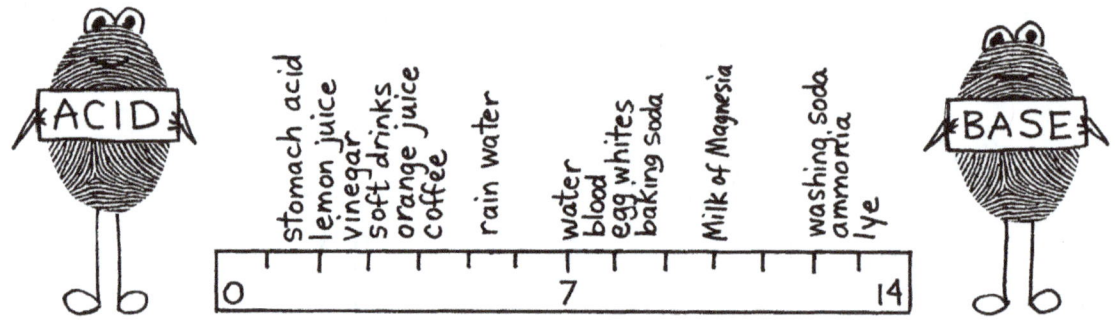

Here's a thought: If acidic substances have lots of hydrogen ions (H^+) and alkaline substances (bases) have lots of hydroxide ions (OH^-), what would happen if you mixed them together? Would all the hydrogen ions decide to attach to hydroxide ions? Yes, that's exactly what happens! And what do a hydrogen ion and a hydroxide ion make when they join together? A normal water molecule. If you put an acid and a base together, the hydrogen ions and hydroxide ions will **neutralize** each other, producing water molecules. Also, in the midst of this reaction, the other pieces of the molecules join together to form a salt compound. Table salt is only one kind of salt, just like table sugar is only one kind of sugar. There's a big family of related compounds that are all called salts. **When you mix and acid and a base, you get water and a "salt."**

Many of us have mixed vinegar (an acid) with baking soda (a base) and witnessed the intense bubbling effect (carbon dioxide being given off). But most of us have never thought about the other product that is produced, a "salt" called **sodium acetate**. You don't normally see this salt because it stays dissolved in the solution. However, if you boil the solution (after all the excitement of the bubbles is over!) to get rid of excess water, you will be left with a solution so rich in sodium acetate that it will begin to form crystals. If you pour out the solution quickly, it will appear to be "freezing" into crystals within seconds. Because the sodium acetate crystals look similar to ice, and because this reaction releases a lot of heat energy, this experiment is often called the "hot ice" experiment.

If acids and bases make salts, is there an acid/base combo that can make table salt, NaCl? Yes, but making NaCl requires chemicals that are not edible, HCl and NaOH, so it's beyond the scope of kitchen chemistry.

Let's take a look at your glass of milk before your salad arrives. We'll set our viewer's magnification on "regular microscope." If you could look at milk through a microscope in a biology lab, this is what you would see.

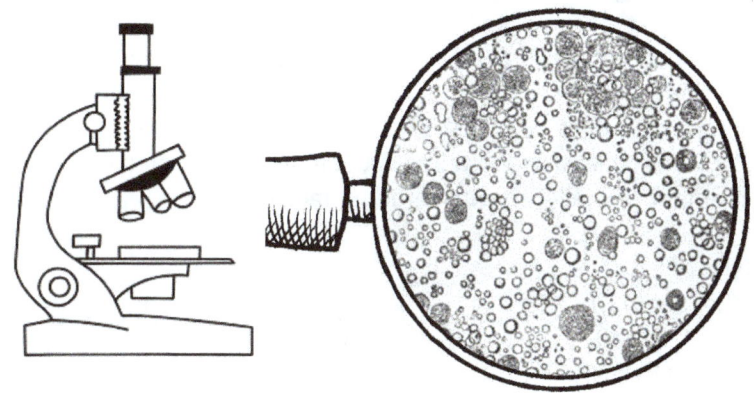

Those little round balls are blobs of fat. They're really small, about the size of a bacteria. And speaking of bacteria, if your milk had not been **pasteurized** (heated) at the dairy it came from, you would have seen bacteria floating amidst the fat blobs. Dairies that sell milk to the general public are required to heat the milk to a certain temperature for a certain amount of time, so that all bacteria will be killed. Pasteurization does a very good job of killing bad bacteria and keeping milk safe to drink. However, it also kills good bacteria, too. Most bacteria are harmless to people, and there even are species of bacteria that you can't live without. Your intestines are filled with "good" bacteria.

It's bacterial warfare all the time in your digestive system as the "good" bacteria try to keep the "bad" ones out. For customers who want these good bacteria in their milk, dairies often sell a type of milk that has had some of the good bacteria put back into it. The most well-known of these good bacteria is **Lactobacillus**. *(lack-to-ba-SILL-us)* The "lacto" on the front of the word means "milk." If you see a dairy product with a label that says, "Contains live cultures," that usually means it has *Lactobacillus* in it. (The most common kind of *Lactobacillus* found in milk is called *Lactobacillus acidophilus*.)

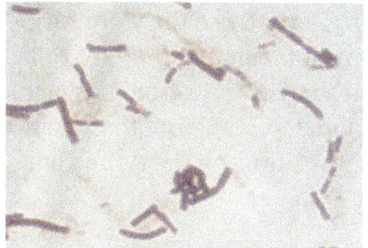

Lactobacillus acidophilus

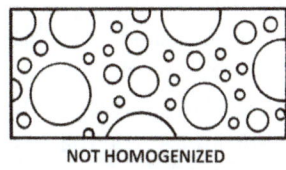
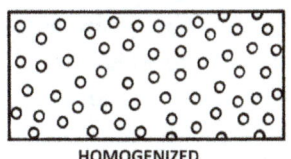

In addition to pasteurization, milk is usually **homogenized** (hom-odge-en-ized). "Homo" means "same" and "gen" means "to make" so in homogenization you are making something the same. If you look at milk that just came out of a cow ("raw" milk) you'll find that the fat blobs are not all the same size. If the milk sits for a while, the larger fat blobs rise to the surface because fat is less dense than water. The fat blobs that collect at the top are known as cream. Farmers who drink raw milk just give their milk jug a good shake to mix the cream back into the milk before they pour it into their glass. However, most consumers don't want to do this. So the dairy presses the milk through a screen with very small holes in order to break the fat blobs into very tiny blobs that are too small to float to the surface. So in homogenization, fat blobs are made to be the same size.

Another process can take some of the fat out of the milk, creating low-fat varieties of milk, such as 2%, 1%, or skim (no fat) milk. Some dairy scientists claim that "raw" (unprocessed) milk is better for your health, but others say store milk is just fine. If milk is not pasteurized, the dairy has to be very careful to monitor the number of microorganisms in it. Rarely, people do get sick from drinking raw milk.

Let's take a closer look at one of those balls of fat. We'll have to switch to our super close-up view where you can see atoms and molecules.

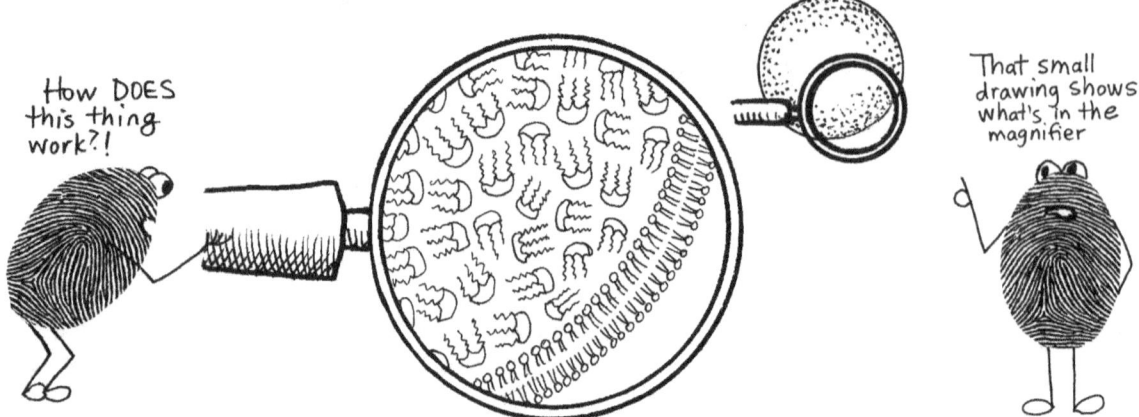

We are looking at just one part of a fat blob so that we can zoom in enough to be able to see its molecular structure. The outer layer, or "shell," of the ball is a very thin membrane, the same kind of membrane that surrounds each cell in your body. The membrane was made by the cells inside the cow's mammary glands. It's only two molecules thick. Those things that look like balls with two tails are the membrane molecules. Inside the membrane are thousands of fat molecules that look like jellyfish with three tentacles.

Let's zoom in on one fat molecule. This type of fat molecule is called a **triglyceride** *(tri-GLISS-er-ide)*. The prefix "tri" means "three." The ending "glyceride" refers to that 3-carbon structure inside the dotted line, **glycerol** *(GLISS-er-ol)*. Glycerol is like a handle that holds on to three very long molecules called **fatty acids**. The "fatty" part of the name comes from the long chains of carbon atoms. All forms or fat and grease are made of long chains of carbon atoms that have hydrogens attached to them. Since the word "acid" is also part of the name, this must mean that they are capable of donating hydrogen ions. It's not obvious from this picture where the hydrogens would come from. Before these tails were attached to the glycerol, there was a hydrogen stuck to the oxygen that is now sitting on the dotted line. That hydrogen comes off as the tail attaches to the glycerol.

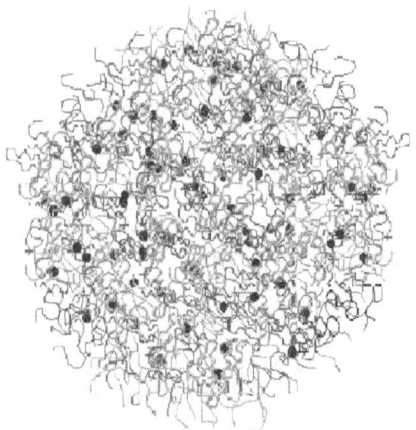

*Little clumps are called **micelles**. This word is used for other types of clumps, too.*

Let's keep going and see what else is in this milk. There are smaller blobs floating around. Let's zoom in on one of them.

There's no membrane around this ball. It looks like a clump of spaghetti and meatballs. The "meatballs" are made of a mineral called **calcium phosphate**. You've probably been told that milk has lots of calcium in it. This is where the calcium is located—it's found in these "mineral meatballs." Milk is a good source of phosphorus, not just calcium. During our discussion of the phosphoric acid in colas, you may have gotten the impression that phosphorus is bad for you. Phosphorus is actually an essential mineral that your body can't live without. Not only is it a main ingredient in bones and teeth, it is also an important part of the ATP molecule that provides energy to all your cells. Here, we see phosphorus working with calcium to keep these protein strands together.

Let's zoom in closer on the "spaghetti noodles."

The "noodles" of the micelle aren't noodles, of course! They are long strings of protein called *casein* (kay-seen). People have been extracting casein protein from milk since ancient times. They didn't know the molecular structure of the proteins, but they knew how to get them out of the milk and use them for paint and glue. Casein paint was commonly used until the 1960s when acrylic paint was invented. Milk glue was in general use until World War II. Also, casein proteins are the basis for many cheeses. (The word "casein" comes from the Greek word for cheese.) Casein has even been used to make a hard "plastic."

Casein paint was used for centuries.

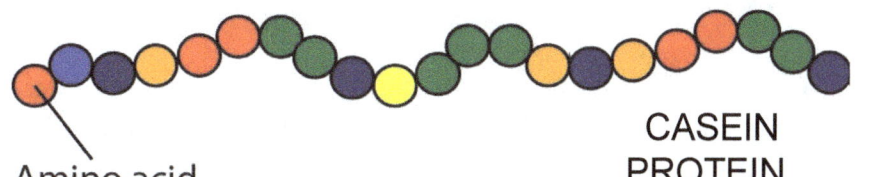

CASEIN PROTEIN

Amino acid

If we zoom in to look at the casein, it might at first look like long strings of beads. Each bead is called an *amino acid.* Amino acids are the individual pieces that make *proteins* like casein. You are already familiar with the word protein. You probably have been told you should eat meat or eggs or beans because they contain protein. Your digestive system tears apart the protein chains until they are single units called amino acids. The digested amino acids will be used by your cells to build and maintain body parts.

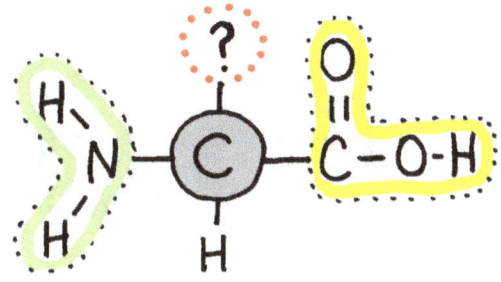

Let's use the highest power on our zoom lens and take a look at one amino acid. Since the word "acid" is in its name, we should expect to find at least one hydrogen, H, that will be able to leave the molecule in the form of a proton.

Amino acids have a carbon atom at their center. This is called the **alpha carbon**. ("Alpha" is Greek for "A.") Imagine that the alpha carbon has four arms. The lines that represent arms are the bonds that it is making. The alpha carbon's bottom "arm" is attached to one hydrogen atom. The top arm is attached to a "wild card" that could be any one of 20 different molecules. Chemists use the letter "R" instead of our nice question mark. The R stands for "residue," but it is easier to think of it as the Rest of the molecule. More about the R in a minute. Let's look at the alpha carbon's left and right arms.

The alpha carbon's left arm is attached to a COOH. Notice the H on the end. It is sitting next to an electron-hogging oxygen atom. The oxygen atom has a strong pull on the hydrogen's only electron. From the hydrogen's point of view, its electron spends far too much time going around the oxygen atom, so the hydrogen is liable to take off and leave its electron behind. When you have hydrogen ions (protons) taking off and roaming around, then by definition, you've got an acid. The NH_2 side of the molecule (the part circled in green) is called the "amine" group. So now we know why they are called "amino acids." The "amino" is NH_2 and the "acid" is the COOH.

Amino acids are not that acidic, though. They can't be put onto the pH scale like vinegar or lemon juice. The H on the end disappears when amino acids hook together to make a chain. The bond between amino acids is called a **peptide bond**. To make this bond, you chop an OH off one side, and an H off another, producing an H_2O.

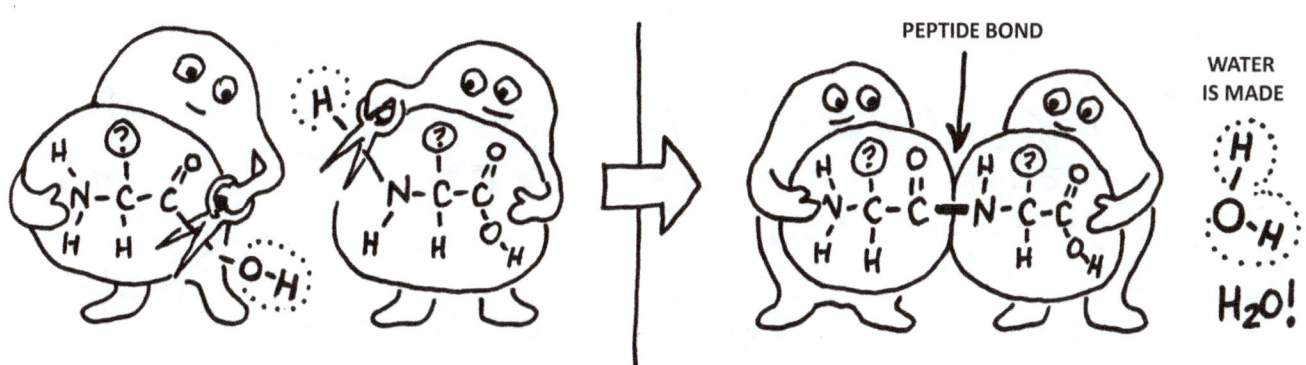

25

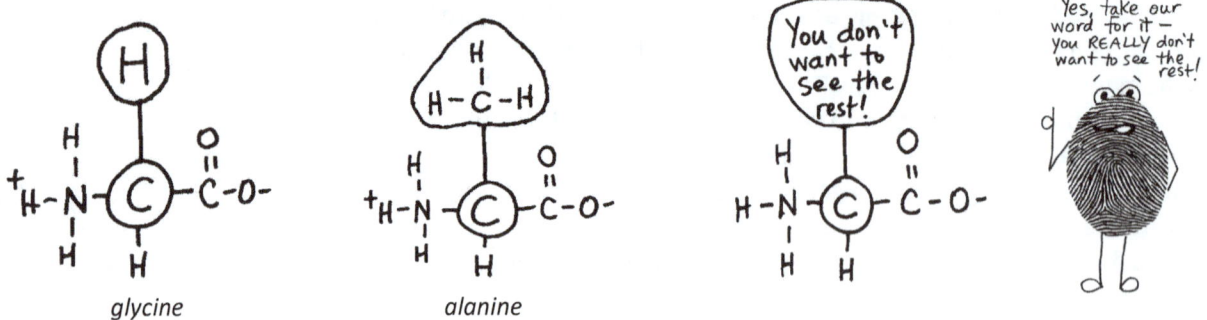

Now just a little bit of info about that "wild card" (?) at the top...

The alpha carbon (the one in the circle) will have one of 20 possible groups of atoms stuck onto that top arm. The simplest possibility is to stick a hydrogen on the end. When it does this, it forms an amino acid called *glycine* (glie-seen). Glycine is the smallest amino acid and is very useful for building things in tight spaces. It is a primary ingredient in collagen, which is found in connective tissue like ligaments and tendons, but it is also abundant in skin and bone. Collagen is like a protein "rope" that is wrapped very tightly, and glycine, because it is small, helps to get the wrap tight. If a carbon and three hydrogens are attached to the top arm, the amino acid *alanine* is formed. If a sulfur is added to that group, *cysteine* (sis-teen) is formed. The remaining 17 possibilities are much more complicated, but they are all based on a unique (one of a kind) arrangement of carbon and hydrogen atoms with an occasional sulfur or nitrogen added in. We'll see amino acids again in a future chapter and learn more about them.

So, back to milk... Casein protein is made of long strings of amino acids. You'll find 18 of the 20 kinds of amino acids in casein. As our enzyme guys demonstrated on the previous page, the amino acids in casein are linked together using **peptide bonds**. Protein chains can have hundreds or thousands of amino acids in them. These long chains are called **polypeptides**. ("Poly" means "many," and "pep" means "protein.")

If we made a model of casein protein using colored beads, we'd have to use 18 different colors!

Enzymes that can break apart peptide bonds are called **peptidases** (pep-tid-ace-ez). The ending "**ase**" is almost always used for an enzyme that breaks things apart. Interestingly, there are several kinds of peptidase enzymes. Some peptidases can only break off the amino acids that are on the very ends of the chain. Other peptidases can get in between the amino acids in the middle of the chain and break them apart. Some can only separate certain kinds of amino acids. Enzymes are highly specialized. As a general rule, enzymes are designed to do only one job.

When peptidases break apart amino acids, they use water molecules to patch the unhappy broken bonds they leave behind, just like we saw in the case of sucrose being broken apart. A water molecule can be split into H^+ and OH^-, and each of these parts can be used as a "band aid" on one side of the broken bond.

This guy is going to use a broken water molecule to patch the ragged ends.

Peptidase enzyme guys tearing apart the amino acids in a protein molecule

Let's take one more look at the milk under our Sooper Dooper magnifier and see if we can find anything else. You can see the edges of the large fat globules with their triglycerides inside, and there's those casein "spaghetti" clumps with their calcium phosphate mineral "meatballs." The tiniest dots are water molecules. But there are some larger dots that look like they might be double rings of some kind. Let's zoom in closer.

Yes, we've found some double-ring molecules. Could they be sucrose? Is there sucrose in milk? Very careful examination of the rings reveals that one of them is glucose but the other is not fructose. The other ring is a type of simple sugar that we have not seen yet. It's almost identical to glucose except that the H and OH on one side are reversed. Seems like a small difference that shouldn't matter at all, but in fact it changes glucose into *galactose*. The existence of galactose was first discovered by the famous scientist Louis Pasteur in 1856. He named his newly discovered chemical "lactose" because it was in milk, but he did not know its molecular structure. Later, chemists figured out the structure and decided to use the word *lactose* to describe the larger two-ring structure, and created a new name, "ga-lactose," for the single ring.

To tear apart the double-ring lactose molecule you need (no surprise) a special enzyme that can snip the bond between glucose and galactose. That enzyme is called *lactase*. Babies of all mammals produce lactase in their digestive systems to that they can digest their mother's milk. In the vast majority of cases, mammals lose the ability to produce lactase as they get older. Not being babies anymore, they don't need to drink their mother's milk. It makes sense. This happens in most humans, too. However, in western Europe many centuries ago, a genetic mutation occurred. The genetic "switch" in the DNA that is supposed to turn off lactase production became broken. Without any instructions to stop, these people's guts go right on producing lactase as if they are still babies. This genetic mistake became very widespread and millions of people today who have European ancestors can drink milk into adulthood. (There are a few places in Africa, also, where some of the population can drink milk.) The ability to drink milk came to be seen as "normal" and therefore people who could not drink milk were considered the defective ones. In modern times, we call this inability to digest milk "lactose intolerance." (Perhaps we should switch the labeling, though, and call the milk drinkers "lactose tolerant," since they are the ones with the broken DNA!) People with lactose intolerance can often take lactase pills that will allow them to digest milk. Cheese and butter are usually less of a problem because much of the lactose has been removed.

Not surprisingly, the milk-drinking Europeans began raising herds of dairy cows to supply them with plenty of milk. They discovered that by controlling the breeding of the cows, they could create cows that could give even more milk per day. A modern dairy cow can give up to 8 gallons of milk every day. That's a lot of milk!

Holsteins are the most popular dairy cow in the world right now.

Holsteins are named after the place they came from in Holland.

Jerseys came from the island of Jersey in the English Channel.

Jerseys are the second most popular dairy breed and are a little smaller.

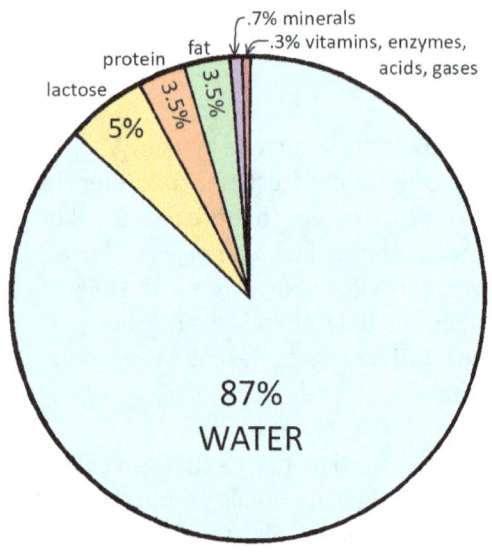

This chart shows you an overview of what cows' milk is made of. (The numbers are averages, so different breeds of cows might have slightly different numbers.) Most of milk is water. For cows, lactose sugar is the most abundant solid substance, followed by fat and protein. Casein is by far the most abundant type of protein, but there are some other minor proteins, too. **Whey** (whay) **protein** is the general term for all these other smaller proteins. They include some proteins that are made by the immune system to fight germs. Others act like "taxi cabs" for transporting things like minerals. There are also are a number of different enzymes and hormones, plus some cow body proteins that leak in accidentally.

We saw the most abundant minerals in milk when we looked at those mineral "meatballs" in the casein protein. Milk also has a small amount of a few other minerals such as iron and zinc. The only category we haven't mentioned at all is vitamins. Milk has most of the major vitamins: A, B, C, D and E. We'll discuss vitamins more in aa future chapter.

This chart is only for cows. The composition of milk depends on what type of mammal is making it. Each animal has milk suited to what the baby needs in that environment. For example, animals that live in cold climates will have a lot more fat in their milk. Animals who nurse their babies for a long time will have milk lower in fat.

One final bit of chemistry before we go on to our appetizers. We need to learn the correct name for the type of liquid that milk is, because milk isn't a solution. Solutions occur when the dissolved particles are extremely small. In salt water, for example, the solute (salt) is made of individual atoms (ions). In sugar water, the sugar molecules are also very tiny. In milk, we have many different types of particles. Some, like lactose, are small. Others, like the fat globules, are large, and can even be seen under a regular microscope. Liquids that have large particles floating in them are called **colloids**. This word comes from the Greek word "colla" meaning "glue." As we learned a few pages ago, casein protein in milk can become sticky and usable as glue. People have been making glue from milk for hundreds of years.

An easy way to determine if a liquid is a solution or a colloid is to shine a flashlight or laser pointer through it. In a solution, the particles are very small and the light will pass right through. In colloids, the particles are large enough that they reflect rays of light that hit them, so the beam of light will be visible. This is called the **Tyndall effect**.

But why is milk white? The answer is as much about light as about milk. Natural light contains all the colors of the rainbow. If an object reflects back all the colors, it looks white. Each tiny particle of fat or protein in the milk is reflecting all of the light that hits it, so the milk looks white. Skim milk, which has had the fat removed, will reflect blue light a little more than the other colors, so it can look slightly blue.

Comprehension self-check

See if you can answer these questions. If not, go back into the chapter and find the information.

1) How does water dissolve something? (What do the water molecules do?)

2) Fat is not a polar molecule. Would it dissolve in water?

3) To encourage carbon dioxide to dissolve into water, should the water be hot or cold?

4) In carbon dioxide, which element gets the electrons more of the time—oxygen or carbon?

5) When molecules begin to move faster, does their temperature go up or down?

6) Which senses flavor, the tongue or the nose?

7) Is benzoic acid harmful to humans? To microorganisms? Why is the FDA concerned about benzoic acid?

8) Which do consumers get more upset about—preservatives in their food, or microorganisms in their food?

9) Why do food companies put phosphoric acid into cola drink?

10) Another name for a hydrogen ion is a _____.

11) When a water molecule breaks apart, what is the OH part called?

12) If a substance has too many protons, is it an acid or a base?

13) Is baking soda acidic or basic?

14) What is the other word that means "basic"?

15) What number is neutral on the pH scale?

16) When an acid and a base combine, they produce _____ and a _____. (page 22)

17) What happens when milk is pasteurized?

18) What happens when milk is homogenized?

19) The most abundant protein in milk is called _____. The strands form a clump called a m_____.

20) What kind of enzymes takes apart proteins?

21) How many amino acids are there?

22) Which part is the "amine" end of an amino acid– the NH_2 end or the COOH end?

23) Lactose is made of what two simple sugars?

24) Two-ring sugars are called disaccharides. Can you name another one, besides lactose?

25) The Tyndall effect is used to tell the difference between a _____ and a _____. (Milk is which?)

> DONT' FORGET about the supplemental videos for this chapter on the "Dissect Your Dinner" playlist at YouTube.com/TheBasementWorkshop

ACTIVITY 2.1   Root beer float word puzzle

"Floats" are a combination of the two things we learned about in this chapter: carbonation and milk (in the form of ice cream). Fill in the correct answers below, then transfer the letters to their places on the float puzzle.

1) The things that are dissolved in solvents. __ __ __ __ __ __ __
   14 65 25 16 11

2) Attraction between water molecules is called __ __ __ __ __ __ __ __ __ __ __ __ __
   77 84 5 31 8 75

3) Plants make caffeine as a __ __ __ __ __ __ __ __ __ (a chemical to kill insect pests).
   68 28 55 81

4) The number of carbon atoms in a glucose molecule: __
   21

5) The number of tastes your tongue can sense: __
   34

6) This ring molecule has this formula: C_6H_6. __ __ __ __ __ __
   67 57 52 95

7) Unprocessed milk is "__ __ __."
   135 46 112

8) Substances that might cause cancer are called __ __ __ __ __ __ __
   10 96 129 12 24 130 133

9) This element has the symbol K: __ __ __ __ __ __ __ __
   50 78 36 115 86 71 18 124

10) Benzoic acid will turn into benzene if it comes into contact with __ __ __ __ __ __
    26 9 2 94 85

11) This acid, found in cola drinks, doesn't dissolve nails! __ __ __ __ __ __ __ __ acid
    138 49 56 82 30 97

12) The correct name for OH⁻ is the __ __ __ __ __ __ __ __
    79 69 60 61 91 29

13) When you combine an acid and a base you get __ __ __ __ __ and a __ __ __ __.
    23 15 140 41 54 20

14) Things that are 7 to 14 on the pH scale are described as __ __ __ __ __ __ __
    58 142 43 93 37 109 27

15) When milk has been heated to kill germs we say that it's been __ __ __ __ __ __ __ __ __
    7 100 51 22 83 70 126

16) When milk has been pressed through a screen it's been __ __ __ __ __ __ __ __
    4 89 48 17 125 64 72

17) The smallest amino acid is called __ __ __ __ __ __
    35 38 98 40 13

18) In a triglyceride fat molecule, the fatty acid "tails" are attached to __ __ __ __ __ __
    59 114 80 121 39 62

19) Strings of casein protein are clumped together in little balls called __ __ __ __ __ __ __ (page 24)
    1 127 32 88 3

20) Enzymes that can tear apart proteins are called __ __ __ __ __ __
    131 33 6 99 19 117

21) The shape of a molecule determines its __ __ __ __ __ __ __ (page 18)
    53 90 128 87 110

22) This agency regulates food and drugs in America. __ __ __
    120

23) Glucose, fructose and galactose are __ __ __ __ __ __ __ __ __.
    143 92 44 113 111 102

24) Triglycerides are made of three __ __ __ __ __ __ __ __ __.
    74 116 45 103 108 63

25) This molecule can be used to patch the unhappy ends of broken bonds. __ __ __ __
    66 42 104

26) Carbonation is an example of a __ __ (CO_2), dissolved into a __ __ __ __ __ (water).
    105 73 106 141 47 107

27) Industrial carbonating machines use high __ __ __ __ __ __ to push the CO_2 into the cold water.
    118 101

28) (We need more A's!) Woodstock, Ontario, is the dairy farming capital of the country of __ __ __ __
    122 134 144

29) (We still need more A's, and another S!) These yellow fruits grow in bunches. __ __ __ __ __
    137 139 119 136

30) (More A's and another S!) Beverages are served in __ __ __ __ __ __.
    105 123 73 76

30

INTERESTING FACTS ABOUT ICE CREAM AND ROOT BEER

1) In the early days of television, this substance was used used in place of ice cream because it wouldn't melt in the hot lights of the studio set. __ __ __ __ __ __ __ __ __ __ __ __ __ __
   1 2 3 4 5 6   7 8 9 10 11 12 13 14

2) America's National Root Beer Float Day is ___ U ___ ___ ___ ___ ___.
   15 16 17 18 19 20 21

3) It takes this many gallons of milk to make one gallon of ice cream: ___ ___ ___ ___ ___ ___
   22 23 24 25 26 27

4) This frozen dessert is sold alongside ice cream, but contains no milk or cream. ___ ___ ___ ___ ___ ___
   28 29 30 31 32 33

5) On average, every American will eat this much ice cream in a year: ___ ___ ___ ___ ___ ___ ___ ___
   34 35 36 37 38 39 40 41

6) The native Yupik people of Alaska make their own version of ice cream. It is called ___ ___ ___ ___ ___ ___,
   42 43 44 45 46 47
and is made of ___ ___ ___ ___ ___ ___ ___ ___ ___, ___ ___ ___ ___ ___ ___ ___,
   48 49 50 51 52 53 54 55 56 57 58 59 60 61 62
___ ___ ___ ___, ___ ___ ___ ___ ___ ___ ___, and ___ ___ ___ ___.
63 64 65 66 67 68 69 70 71 72 73 74 75 76 77

7) What did Nancy Johnson of Philadephia invent in 1843? ___ ___ ___ ___ ___ ___ ___ ___
   78 79 80 81 82 83 84 85

8) The city where the ice cream cone was invented at the World's Fair in 1904: ___ ___ ___ ___ ___ ___ ___
   86 87 88 89 90 91 92

9) The biggest consumers of ice cream are these countries (in order of consumption):
___ ___ ___ ___ ___ ___ ___, ___ ___ ___ ___ ___ ___ ___ ___, ___ ___ ___ ___ ___ ___
93 94 95 96 97 98 99 100 101 102 103 104 105 106 107 108 109 110 111 112 113 114

10) Root beer was originally made from the roots of this tree. ___ ___ ___ ___ ___ ___ ___ ___ ___
    115 116 117 118 119 120 121 122 123

11) Native North Americans used this tree (in #10) for making ___ ___ ___ ___ ___ ___ ___ ___
    124 125 126 127 128 129 130 131

12) In places where this tree is not available, this plant is used instead because it has a similar flavor to root beer:
___ ___ ___ ___ ___ ___ ___ ___ ___ ___ ___ ___
133 134 135 136 137 138 139 140 141 142 143 144

ACTIVITY 2.2 Mammal milk trivia

All female mammals make milk. The chemistry of each animal's milk is just what its babies need. See if you can match these descriptions with the correct mammals. They aren't easy! Use any clues you can in the descriptions, including geography and animal behavior.

Possible answers: sheep, goat, donkey, whale, seal, horse, black rhino, wallaby, hippo, rabbit

1) The milk of this mammal holds the record for being highest in fat content. The mother only feeds her baby for about a week, but during that time the baby will double its weight, as well as putting on a thick layer of fat under the skin. The baby will need a lot of fat to protect it from the cold. _____

2) The milk of this endangered mammal holds the record for being lowest in fat content. The mother's body can't put a lot of energy into the fat content of the milk because her pregnancy lasted for over a year, and then she will nurse her baby for over two years. _____

3) The milk of this mammal is excellent for making cheese because it is high in both fat and protein. (It has twice the fat content of cow's milk.) Cheeses often made with this milk include feta (Greece), Roquefort (France), and ricotta (Italy). This mammal only produces milk naturally at certain times of the year because of seasonal breeding. To get year-round milk production, farmers must give hormone shots to these animals. _____

4) This mammal produces milk that has one of the highest protein levels in the animal kingdom. The milk also has twice as much fat as cow's milk. The mother only nurses her babies in the morning and evening and spends all day foraging for food (in gardens if she gets the chance). Mammals that only nurse their babies once or twice a day often have milk that is high in protein and fat. Those few meals have to be good ones! _____

5) This mammal's milk is the subject of an untrue "fact" that circulates on the Internet. The Internet rumor says that this mammal's milk is pink. Supposedly the milk mixes with a red body chemical, and the red and white combine to make pink. The part about the red chemical is mostly true, although it is clear when it is secreted by the skin. This chemical acts as a natural sun screen, turning red and then brown as it absorbs UV rays. The milk produced by this mammal is white, as is the milk of every mammal on the planet, although direct studies of the milk have rarely been done because of the ferocity of the animal. It would be very hard to get close enough to a nursing mother without being injured or killed. _____

6) The milk of this mammal does not separate into milk and cream. The fat globules are bound to the other solids in the milk so they are not able to float to the top. Milk experts say that this is one of the most digestible milks and one of the most similar milks to human milk. It is often used to make cheeses. The milk has a strong flavor to it (tasting a bit like the animal smells), which makes it less popular than cow's milk. _____

7) It is critical that this mammal's milk be high in fat so that the milk won't mix with the water around it. If the milk was low in fat it could more easily mix with water, making it difficult for the baby to get enough of it into its mouth. The mother's teats are not visible most of the time and only come out when the baby nudges them. _____

8) This mammal's milk was first recommended by Hippocrates in 400 BC. In the ancient world it was used both as a health remedy for sick infants and as a skin cosmetic product for women. Right up until modern times this milk has been used to feed orphaned human babies if no source of human milk was available. The nutritional content of this milk is very similar to human milk except that it is slightly lower in fat. The babies would be given liquid fats such as olive oil to make up for this difference. _____

9) In central Asia and Mongolia, the milk of this animal is used to make a fermented drink called kumis. _____

10) This mammal can produce different types of milk in different teats because she can have babies of different ages both suckling at the same time. The teats that are suckled by the tiny infant in her pouch will produce milk high in sugar. The teats for the older babies will produce milk low in sugar but high in fat and protein. _____

ACTIVITY 2.3 Second installment of "Chew It Over," a group game to be played during a meal

Here is another round of questions for you to use at a mealtime that you share with family or friends. These questions relate to the topics we learned about in this chapter. Again, you can use these questions in a variety of ways. You can be the quiz master and determine who gets which questions, or you can cut the questions out of the book and put them into a bag or bowl and let people choose a question randomly. The answers on are the back of this page.

CHAPTER 2: CARBONATED BEVERAGES and MILK

1) The average cow can produce about how many glasses of milk each day?

CHAPTER 2: CARBONATED BEVERAGES and MILK

2) In the U.S., which month is National Dairy month?

CHAPTER 2: CARBONATED BEVERAGES and MILK

3) There are some pretty strange soft drink flavors around the world. All of these are real flavors except one. Which one is not a real flavor?
a) Black Garlic	b) Onion
c) White Fungus	d) Mustard

CHAPTER 2: CARBONATED BEVERAGES and MILK

4) Until 1950, this carbonated drink contained lithium citrate, which is today used as a brain medicine:

a) Coke	b) Pepsi	c) 7-Up	d) Dr Pepper

CHAPTER 2: CARBONATED BEVERAGES and MILK

5) About how many teaspoons of sugar are in a can of soda (pop/coke)?

CHAPTER 2: CARBONATED BEVERAGES and MILK

6) Which of these frozen desserts does not contain milk?
   a) sherbet (sherbert)	b) sorbet
   c) spumoni	d) gelato

CHAPTER 2: CARBONATED BEVERAGES and MILK

7) The name of the orange-flavored soft drink "Fanta" is a German word for what?

a) imagination	b) happiness
c) intelligence	d) courage

CHAPTER 2: CARBONATED BEVERAGES and MILK

8) Years ago, there was a rumor that ship-wrecked sailors from France used the milk of one of the native animals on the island to make some cheese. Can you guess the animal?
a) mice	b) rats	c) pigs	d) rabbits

CHAPTER 2: CARBONATED BEVERAGES and MILK

9) Have you ever tasted goat milk or goat cheese? Would you recommend it?

CHAPTER 2: CARBONATED BEVERAGES and MILK

FUNNY FACT: When Pepsi's slogan "Come alive with Pepsi" was translated into Chinese, it said, "Pepsi brings your ancestors back from the grave."

CHAPTER 2: CARBONATED BEVERAGES and MILK

11) What is your favorite carbonated beverage?

CHAPTER 2: CARBONATED BEVERAGES and MILK

12) Which is your least favorite carbonated beverage?

1) About 100
2) June
3) Mustard
4) 7-Up
5) 10 teaspoon
6) sorbet
7) imagination
8) rats

3: Appetizers

That's okay. There is plenty here to dissect! It will take a whole chapter to get through all the chemistry found in butter, cheese, crackers and bread.

Good idea. Let's start with the butter. Since butter is made from milk we should see lots of familiar things.

The word "butter" is a very old word. It can be traced back to the Latin word for butter, "butyrum," which came from the ancient Greek word "bouturon" meaning "cow cheese." ("Bous" was "cow," and "turos" was "cheese.") We also saw that the word "casein" came from a Greek word for cheese. There were many kinds of cheese back then, just like there are now, so it's not surprising to find more than one word for cheese. Without refrigeration, milk was hard to keep. The natural bacteria found in raw milk made it spoil within a day or two. Ancient peoples found that turning the milk into butter, cheese or yogurt made the milk stay edible for weeks or even months. When butter does finally spoil, we say it goes "rancid." This is due to a smelly acid substance produced by bacteria living in the butter.

Let's get out that Sooper Dooper magnifier and take a look at our butter.

Do you recognize those things that look like stretched out letter E's? (We called them weird-looking jellyfish in the last chapter.) Most of them have three "tails." They're triglycerides. The "tri" part means "three" and the "glyceride" part refers to the hanger that the tails are attached to. The tails are called fatty acids and they are made of long strings of carbon atoms with hydrogens attached. They are the smallest type of fat. When we looked at milk, we saw triglycerides inside those big, round globules. Here, the triglycerides are just scattered about everywhere, not inside globules. What happened?

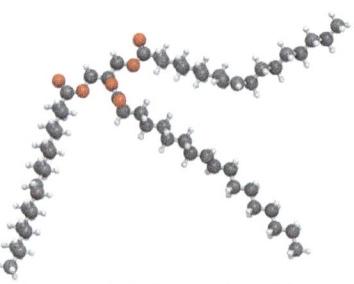

a model of a tryglyceride

The process of churning butter is all about destroying those fat globules. You bang and smash those globules around as if they were microscopic piñatas filled with fats instead of candy. Once they are all broken open and the triglycerides are no longer contained inside the globules, the fats stick together to form one solid mass. That's butter.

As the fats stick together, most of the water that was between the fat globs gets squeezed out. Part of the butter-making process includes draining off water. This water won't be pure water, however, but will still have a little bit of fat, protein and sugar in it. Dairies usually save this drained off water and either sell it as "buttermilk" or use it as an ingredient in other products such as ice cream.

In the past, buttermilk was used as a source of acid to either curdle milk for cheese, or to react with baking soda in biscuit and bread recipes. But wait... milk as a source of acid? Fresh milk is definitely not sour tasting. How can a milk product be a source of acid? The answer lies in the label on the buttermilk carton. It might say "cultured" buttermilk. In food science, "culture" doesn't refer to great art and literature. A "culture" is a source of microorganisms, often bacteria. So "cultured" buttermilk means it contains bacteria.

These milk bacteria are not harmful like the ones that give you strep throat or pneumonia. Most of the bacteria found in milk are very good for us. We want them to live inside us because they fight against any bad bacteria that might get into our intestines. But even good bacteria need to eat, and these **Lactobacillus** *(LACK-toe-ba-SILL-us)* bacteria eat the lactose sugars in the milk. As part of their digestion process, the bacteria produce a chemical called **lactic acid**. This acid isn't nearly as acidic as lemon juice or vinegar, but it's acidic enough to cause milk to **curdle** (form solids). Nowadays, most milk is pasteurized, so dairies must add a bacterial culture to their buttermilk in order to sell it as "cultured" buttermilk. Some cooks prefer to make instant buttermilk in their kitchen by adding lemon juice to fresh milk.

Let's look closer at the magnified butter. What are those other blobs and dots, in and around the triglycerides? A few of them look like tiny bits of protein—chains of amino acids. They could be little pieces of casein that broke off the micelles, or they could be some of those smaller milk proteins, the miscellaneous ones that came from the mother cow. There aren't very many of them, so butter isn't a very good source of protein.

The tiniest dots might represent lactose sugars, but just a few of them. Most of the lactose molecules would have stayed down with the liquid milk and not floated to the top with the cream. When the cream was taken off the top, most of the lactose stayed behind. Butter has a small enough amount of lactose that people who can't digest lactose can often still eat butter.

The rest of the dots represent molecules that give butter its yellow color: **beta-carotene**.

The dots are carbon atoms. There are also many hydrogen atoms attached to the carbons, but all chemists know that the H's are there, and they get lazy and don't bother drawing them. The double lines represent places where carbon atoms are double-sharing their electrons. Notice the pattern of the bonds across the middle of the molecule: double, single, double, single, double, single, etc. This is important. Molecules that have this pattern often reflect light. In this case, beta-carotene *reflects* orange and yellow light. Of greater importance is the fact that it *absorbs* violet, blue and green light. Beta-carotene is part of the light-collecting system in leaves. The light that is collected is used in the process of photosynthesis, where sunlight, CO_2 and water are turned into glucose sugar.

You might have wondered if the name "carotene" has anything to do with carrots. Yes! Carotenes were named after carrots. As a general rule, orange or yellowish-orange colors in plants are caused by the presence of beta-carotene molecules. Other sources of beta-carotene include cantaloupes, sweet potatoes, mangoes and pumpkins. But how did these plant molecules get into the butter?

The milk that the butter was made from came from cows who ate leaves that had beta-carotene molecules in them. (Green leaves often contain orange or red pigments, but the green "drowns them out.") The leaves were digested in the cow's stomach, releasing the beta-carotene molecules from the plant cells. The beta-carotene molecules then went floating around inside the cow's body, looking for a place to stay. Since the beta-carotene molecule is rich in carbon atoms, it is naturally attracted to other molecules that have a lot of carbon atoms. Triglycerides have three long strings of carbon atoms, so beta-carotene molecules feel right at home in and among them. A lot of animal fat is light yellow in color, due to the presence of these beta-carotenes. The yolks of chicken eggs are rich in beta-carotenes for the same reason.

As long as the beta-carotenes stay inside the fat globules in milk, the rays of light can't reach them very well. The outsides of the fat globules scatter all the colors of light equally, so the milk looks white. However, when you make butter, you must smash all those globules open. The triglycerides come spilling out, and so do the beta-carotenes. Once the beta-carotenes are out, they begin to reflect yellowish-orange light, so the butter looks light yellow. People who make their own butter from their own cows notice that the color of their butter changes from season to season, depending on what plants the cows are eating. The butter is more or less yellow at certain times of the year.

The cow's body (and your body) hang on to these molecules of beta-carotene so that they can be used to make **vitamin A**. Beta-carotene can be cut in half by a scissor enzyme to make two molecules of vitamin A. Those two red dots are oxygen atoms that will be used to patch the ends.

Here are the two halves of beta-carotene, shown as ball and stick models. Now you can see all the hydrogens, and you can see the two oxygens that the enzymes used to patch the cut ends. Each half is now a molecule of **retinol**, a form of vitamin A. Sometimes other enzymes come over and "tweak" the retinol molecule, making tiny changes that will turn it into different versions of vitamin A.

The most well-known of these variations is called **retinal**. Retinal is used in the cells of your retina, the area at the back of your eye that senses light. Retinal becomes part of a molecule that absorbs photons of light. (This makes sense, since beta-carotene absorbs light in plant leaves.) Retinal sits inside a special holder made of protein. Without retinal, the holder is useless. There are billions of these holders in the cells of the retina and they must all be filled with retinal. If you don't eat enough beta-carotene or vitamin A, the retina won't work right and your vision will be affected. This is why people say carrots are good for your eyes. However, a little beta-carotene goes a long way, and over-eating carrots isn't going to give you super-power vision.

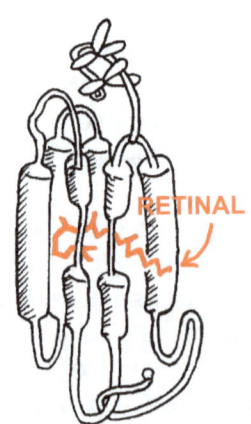

After beta-carotene is chopped in half, the resulting retinols don't reflect as much yellow and orange light. Animals that store retinol (vitamin A) in their fat, instead of beta-carotene, will produce milk that makes white butter instead of yellow. Butter and cheese made from goat and sheep milk is white. Consumers seem to prefer yellow butter over white, so beta-carotene is often added to sheep butter to make it yellow.

Since cheese is made from milk, will we find the same molecules and structures that we found in milk and butter? Let's take a look.

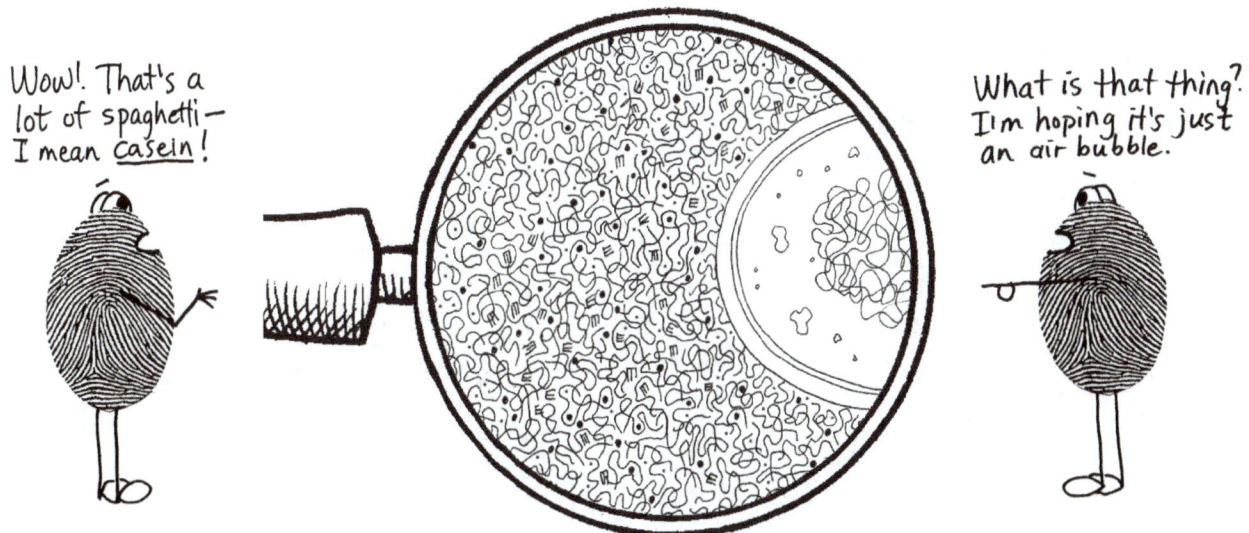

What happened to the casein "spaghetti" balls we saw in the milk on page 24? This is a mess! Casein "noodles" are all over the place! We can see some triglycerides in there, too, so this cheese has some fat in it. Those really tiny dots might be beta-carotene molecules helping to give the cheese its orange color. But what is that HUGE thing sticking into the picture? It looks like part of something that's bigger than our viewing area. We'll have to reduce the magnification a bit in a minute to get a better image of it.

Let's find out what happened to our casein micelles first, then we'll delve into the mystery of that huge whatever-it-is.

So what did happen to the casein micelles? To understand what happened, we need to learn a little more about the micelles.

Imagine that our micelle is made of yarn instead of protein, and let's use blue as our color scheme. We'd need to use four different shades of blue because there are four different types of casein protein in our micelle. We'd wind many small balls of yarn and then stick them together. (The small balls are called sub-micelles.) A real casein micelle might have as many as 500 of these smaller balls. One of these types of casein has long fringes that hang off, making the ball look furry. We can stick little orange beads between the balls to represent those mineral "meatballs" made of calcium phosphate. These calcium phosphate balls act sort of like magnets to keep the micelles together.

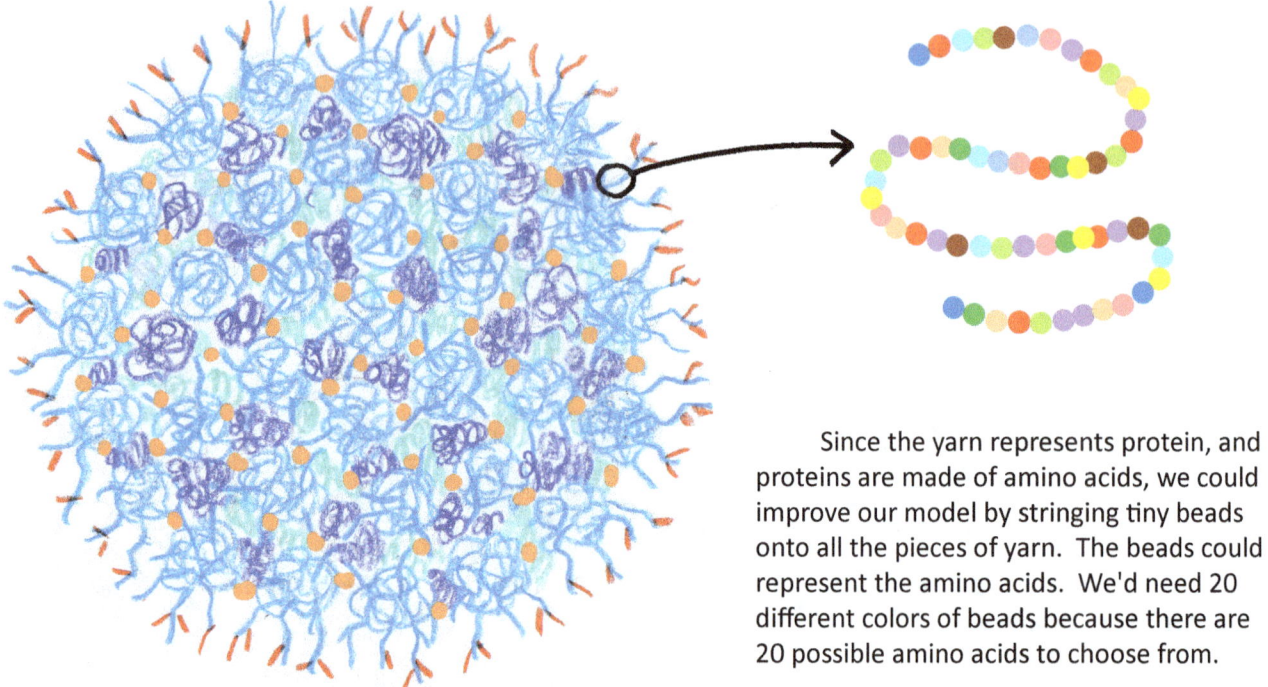

Since the yarn represents protein, and proteins are made of amino acids, we could improve our model by stringing tiny beads onto all the pieces of yarn. The beads could represent the amino acids. We'd need 20 different colors of beads because there are 20 possible amino acids to choose from.

Now, we must remember that amino acids aren't really little balls. Although chemists represent them as circles, you will remember that they are really groups of atoms. There's a carbon atom in the center, with a COOH on one side, and an NH_2 on the other.

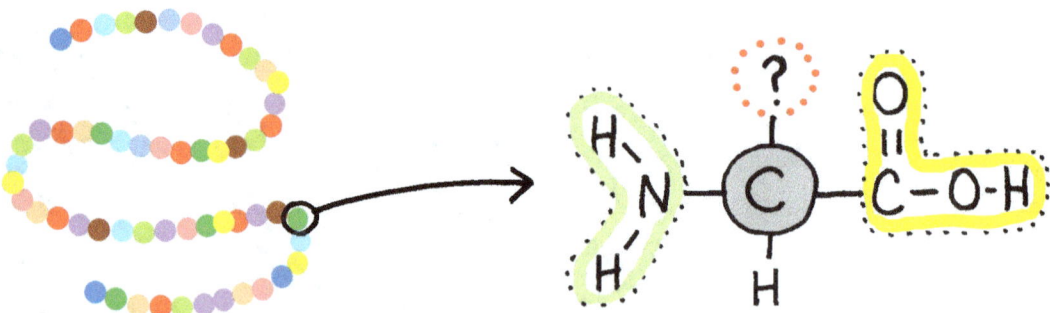

That (?) at the top of the molecule turns out to be very important. That's the part that makes each amino acid different from the others. We saw that glycine, the smallest amino acid, has only an a hydrogen atom, H, in that place. That makes glycine special because it the smallest amino acid. Some amino acids have a chain of carbon atoms hanging off, looking like the fatty acids we saw on the triglycerides. Amino acids that have extra strings of carbon atoms (where the "?" is) hate water and want to hang out with fats. Other amino acids are just the opposite and have extra atoms (where the "?" is) that love to be around water. A few amino acids have an atom of sulfur hanging off, which makes them good at building bridges. We'll meet an amino acid with sulfur when we dissect the bread. Here, in the fringy casein proteins, we find the amino acid **threonine**, which is particularly good at attaching itself to sugar molecules.

When a string of amino acids grabs and hangs on to some sugar molecules the result is a molecule that is a combination of protein and sugar, and we call it a *glycoprotein*. ("Glyco" comes from "glucose" but it can mean any kind of sugar molecule.) Those red strings hanging off the blue yarn are glycoproteins. These glycoprotein molecules are largely responsible for the ability of the casein micelles to float around in the milk without clumping together. Which is a good thing while the milk is in the cow.

The red fringes help to give the casein micelle a negative electrical charge. "Like" charges repel, so the micelles stay away from each other because they are all negative. The same thing would happen if they were all positively charged. But in this case we see negative charges.

The first step in making cheese, yogurt and some milk-based desserts is to get those casein balls to clump together. We call this *curdling* the milk. Depending on how it is done, the milk will either turn lumpy, like cottage cheese, or it will get thick, like yogurt and custard. In cottage cheese, the curds are large and very visible. In yogurt or custard, the curds stay so small that the texture remains smooth and creamy.

There are basically two ways to make these casein micelles clump together (curdle).

1) How To Stop Casein Micelles from Being So Negative

Put some positives into their environment! Why don't we toss in some positive hydrogen ions (protons) and have them go over and cancel out some of the negativity? And where do we get a good supply of protons? From acids! That's the definition of an acid—something that gives away protons. Acids that are right there handy in the kitchen are vinegar and lemon juice. Vinegar is the most common substance used for curdling milk to make cottage cheese and Ricotta cheese.

2) How To Cut Off the Negative Fringes

When you need to cut molecules, what do you use? Enzyme scissor guys, of course. You just need the right little enzyme robot, and he'll go over and snip those fringes right off. Where would we find this particular robot? Well, since baby cows' stomachs are good at digesting cow milk, perhaps this would be a good place to look. And, in fact, we will be successful, because people have been extracting this enzyme from baby cow stomachs for thousands of years. Chemists call it *chymosin*, but food scientists and chefs usually cal it *rennet*. Rennet is the word you will hear more often. The most famous brand name of rennet is Junket®. (NOTE: There are now plant-based sources of rennet, too.)

Junket tablets and rennet liquid drops

2.5) Another Source of Acid: Bacteria

When milk goes sour on its own, you still see curdled white solids and clear whey. In this case, the bacteria that are making the milk go rotten are producing acids. They are eating the lactose sugars and making lactic acid as a waste product.

Some cheeses require just this first step. Put in some acid and let the milk curdle. After you see the white clumps of casein forming, you can strain them out and use them for your cheese. The watery stuff that is left over is called **whey**. Whey still has some protein in it, mainly those smaller proteins that aren't casein. Whey proteins can be taken out and dried into a powder. (Some people are really into eating powdered whey protein, and you'll see it sold in health food stores mainly as a supplement for athletes.) In ricotta cheese, the whey is strained off and only the solids are used. In cottage cheese, the whey can be part of the final product. In commercial cottage cheese, the whey often has a thickening agent added, as people don't tend to like watery whey.

Cottage cheese is made of curds and whey.

Hard cheeses—the ones that come in blocks that you can slice—have had a second ingredient added: microorganisms in the form of either **bacteria** or **mold** (or both). The first cheeses ever made contained the natural bacteria and molds that were in that environment. The microorganisms would have come from places like the skin of people or animals, from plants or dirt, or from barns and houses. Each part of the world had its own unique blend of microorganisms. The spores from these microorganisms would go into the air and float around, and if a pot of milk or curds was left sitting out for a while, it would collect spores that fell out of the air. The bacteria and mold would start to grow in the curds and eventually turn them into a hard cheese.

As time went on, cheese makers discovered that they didn't need to wait for spores to fall out of the air. All they needed to do was to save a bit of the last batch of cheese, and add it to the new batch. They didn't understand that they were saving microorganisms, though, because they didn't have microscopes and had no idea that tiny living things were inhabiting their cheese. They also discovered that cheeses made in different places had different flavors. Often, a cheese would be named after the town where it was first made. Cheddar, for example, started out in the English town of Cheddar. This type of cheese became so popular that people outside of Cheddar wanted to make it. They bought cheese in Cheddar, then went elsewhere and added the Cheddar cheese culture to their own curds. Now, Cheddar cheese is made all over the world, but the original culture came from England.

Scientists can now analyze and identify the exact species of microorganisms found in each type of cheese. Cheddar and Colby have *Lactococcus lactis, Lactobacillus casei, Streptococcus cremoris,* and *Streptococcus durans*. (You don't have to try to pronounce those!) Blue cheese has a blue mold called *Penicillium roqueforti*. Swiss cheese has an unusual bacteria that produces carbon dioxide as a waste gas. The carbon dioxide forms bubbles and gives Swiss cheese its holes. Limburger cheese has a species of bacteria that is extremely similar to the bacteria found on human feet. No wonder Limburger cheese smells like stinky socks!

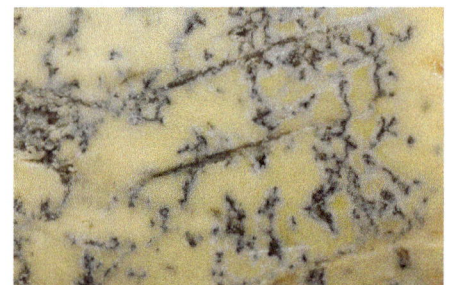

A cheese with blue mold in it.

New cheeses are still being invented. For example, a cheese called Cornish Yarg (that's "Gray" spelled backwards) was first made in Cornwall, England, in 1984. Some Cornish cheese-mongers (that's what you call professional cheese makers) found a very old recipe and decided to try it. They added garlic to their cheese and then wrapped the blocks of cheese in the leaves of the nettle plant. Another new English cheese is called "Stinking Bishop," inspired by the cheese recipes of monks during the Middle Ages who washed their cheeses in pear juice.

Photo credit: Tristan Ferne, UK, from Wikipedia article on Cornish Yarg

We just saw one end of the bacteria.

Bacteria make acid as a waste product when they eat. All living things make waste products as a result of eating. Some of the waste products are obvious, like... well, you know... we flush them. Others are not so obvious, like the carbon dioxide that goes out of your lungs when you exhale. Or the lactic acid produced in your muscles when you push them to their limit. Here in our cheese, the bacteria have been doing exactly the same thing that your tired muscles do; they are producing lactic acid as a result of burning glucose molecules for energy.

Glucose is our body's primary source of energy. There is energy locked up in the bonds that hold the carbon atoms together. If you break those bonds, you can release the energy. The first step in using glucose for energy is cutting it in half. As you might guess, you'll need enzyme scissors to help with this process. However, you won't need just one enzyme, but TEN of them! The glucose must be snipped, twisted and patched until two molecules of **pyruvate** *(pie-RU-vate)* are formed.

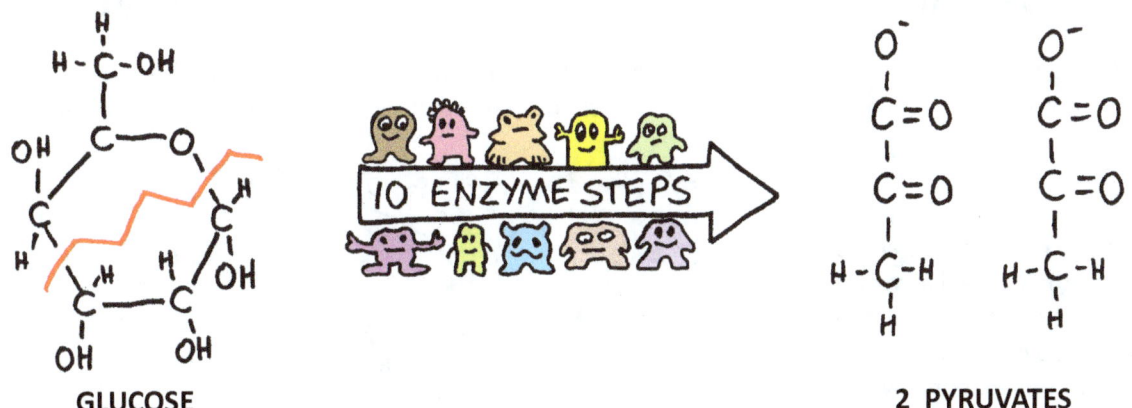

GLUCOSE **2 PYRUVATES**

At this point, animal cells (including yours) send these pyruvates to special organelles inside the cell called the **mitochondria**. Inside the mitochondria, the rest of the bonds between the carbon atoms will be broken to release the remaining energy. Bacteria don't have mitochondria, though, so they can't cut apart those pyruvate molecules. They use them for a different purpose.

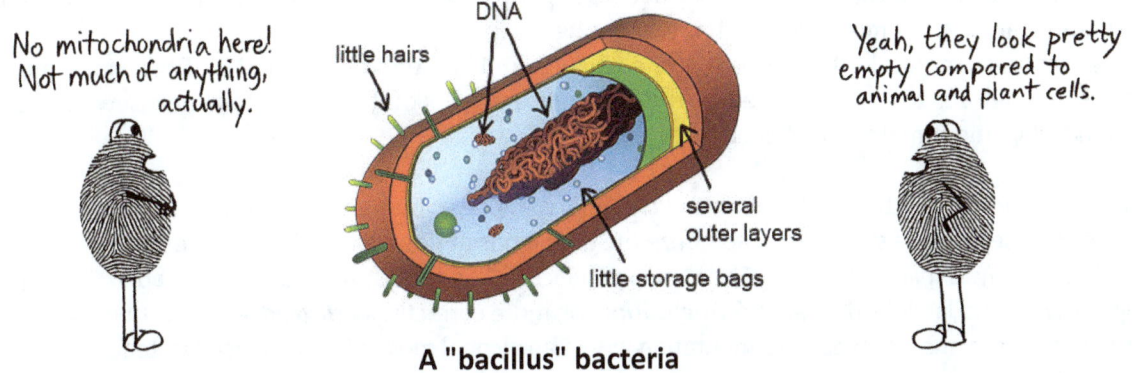

A "bacillus" bacteria

43

The bacteria living in our cheese use the pyruvate molecules to get rid of some atoms are that preventing them from splitting more glucose molecules. We didn't show you all the complicated chemistry going on in those 10 steps. We just drew some cute enzyme guys and let it go at that. But if we showed you every single atom coming and going in this process, you'd see that part of the process involves putting electrons into molecules that act like shuttle buses. The shuttle buses have to be emptied before they can be loaded again. If the shuttle buses are all sitting there full, the process of splitting glucose will stop. So the bacteria have enzyme guys that can unload the shuttles and put their cargo onto the pyruvates. The cargo that is transfered is electrons and protons. The end result is that pyruvate gains 2 hydrogen atoms.

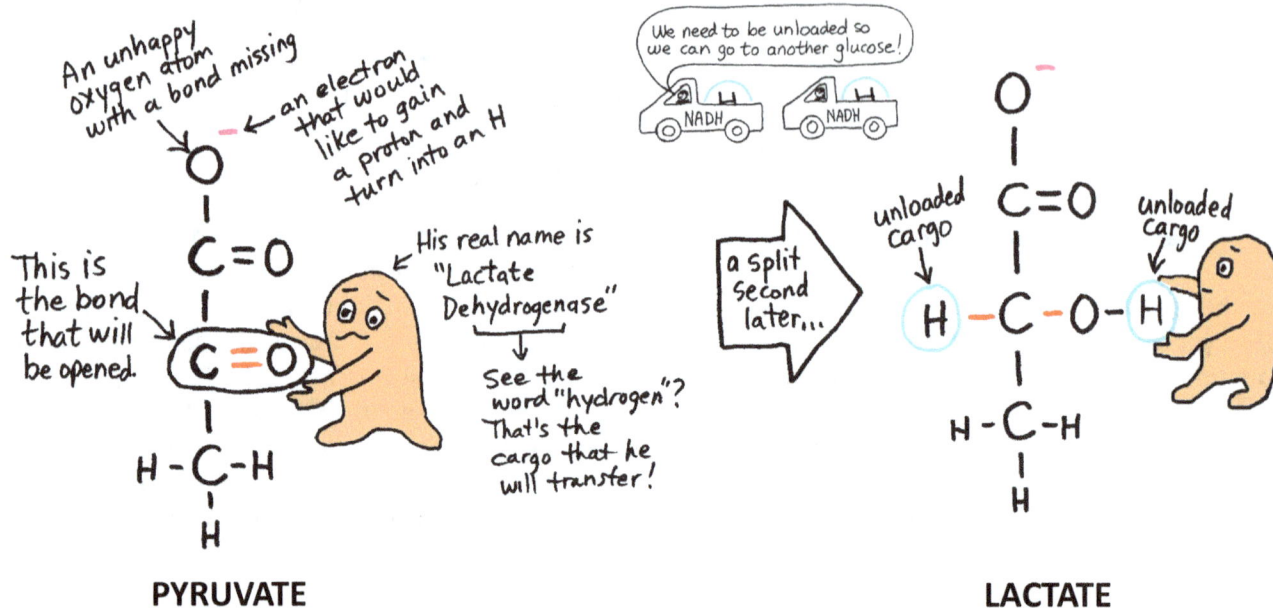

PYRUVATE　　　　　　　　　　　　　　　　　　　　**LACTATE**

With the H cargo unloaded onto pyruvate, those shuttle bus molecules will now be able to help with the splitting of another glucose molecule. Mission accomplished! (The splitting of glucose will release some energy.)

Now.. what about the "acid" part of this process? We keep talking about **lactic acid**, but so far all we have is something called **lactate**.

That lonely pink electron on the O at the top of the molecule is bound to pick up a passing hydrogen ion (proton). The electron and proton will join together to make a hydrogen atom, and then we'll be able to draw an H connected to that O. (The yellow arrow is pointing to this newly formed H.) The molecule will then officially become lactic acid. It's an acid because that proton that just came in could just as easily decide to go wandering off again. Protons change their minds a lot. By definition, an acid is a molecule that has wandering protons.

The lactic acid molecules then go out of the bacteria and into the milk around it. Once in the milk, guess what that proton decides to do... yep, it leaves the molecule and goes off to seek its fortune elsewhere. Loose protons wandering about make a substance taste sour.

Before we turn our attention to bread and crackers, we need to burden you with two vocabulary words. (Most biology courses require students to memorize the definitions of these words.) The process of splitting glucose in half is called **glycolysis** (glie-KOL-i-sis). ("Glyco" means "glucose," and "lysis" means "to break apart.") The process of turning pyruvate into lactic acid is called **fermentation**, or, more correctly, **lactic acid fermentation**. When we dissect bread, we'll see another type of fermentation where an alcohol molecule is produced instead of lactic acid.

What is bread made of? If we look at bread under a magnifier, we'll see something like this. It looks a bit like a sponge. When the dough was first made, it was smooth and dense. The kneading process forced out any air that was in the dough. But when the bread came out of the oven it was fluffy and spongy. What happened? After we find out the answer to this question, we'll use the Super Duper magnifier to find out what the molecules look like.

The tiny air holes in the dough were created by another type of fermentation: **ethanol fermentation**. This fermentation is done by **yeast** instead of bacteria. The yeast cells "eat" the sugars in the bread and produce **carbon dioxide** and **ethanol** as waste products. The carbon dioxide bubbles into the bread dough and produces those holes. The bread hardens as it bakes, so the holes are preserved long after the carbon dioxide is gone.

Ethanol *(ETH-uh-noll)* is a type of alcohol. It is found in alcoholic drinks such as beer and wine, and it can also be used as a fuel. Many gas stations now sell gasoline (petrol) that has ethanol in it. If you see "E10" written on a gas pump that means that the gasoline contains 10% ethanol. In bread, the alcohol evaporates during baking, so the final product doesn't actually contain any alcohol.

The yeast cells produce ethanol for exactly the same reason that bacteria produce lactic acid. Eating sugar always starts out with glycolysis—splitting a glucose molecule in half. Two of those ten steps of glycolysis involve filling a shuttle bus (shown as a pick-up truck) with an H atom. After completing glycolysis they must find a way to empty their little shuttle trucks so that they can fill them again. The bacteria had an enzyme guy

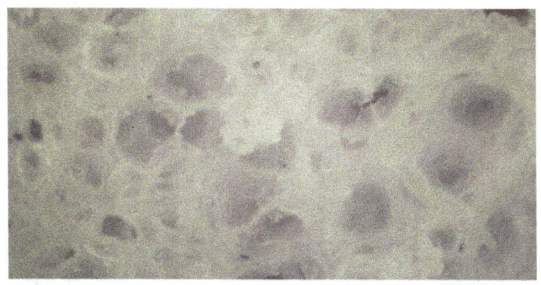

who could open up the double bond on the middle carbon atom and then transfer the two hydrogen atoms from the trucks to the pyruvate molecule, turning it into lactate.

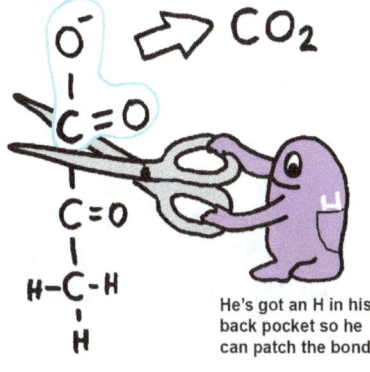

He's got an H in his back pocket so he can patch the bond.

The yeast cells have a similar enzyme guy: **pyruvate decarboxylase**. This name isn't as hard as it looks! "De" means "from," "carboxyl" is the correct name for COOH, and "-ase" is the label for "enzyme." So the name means "the enzyme that takes COOH from pyruvate." However, you'll notice that the COOH has already been the victim of hydrogen unfaithfulness, and is only COO$^-$. The hydrogen went wandering off as nothing but a proton, leaving its electron behind. (We'll be nice and still refer to it as "carboxyl," politely ignoring its recent loss.) When the COO$^-$ is clipped off, it goes floating away as CO_2, carbon dioxide. That's how the bubbles in bread are produced.

Now it's time to make ethanol. Another enzyme comes along and opens the double bond on the carbon. If you look back to page 44, you'll see an orange enzyme doing basically the same thing. In this case, though, our enzyme worker is going to transfer the hydrogen cargo in such a way that the molecule will turn into ethanol.

The ethanol in bread is destroyed as it is baked. When bread comes out of the oven the ethanol is gone. However, in other foods the ethanol is an essential ingredient. In beer and wine, the ethanol stays. Alcoholic drinks can be made from just about any type of plant. Yeast isn't too picky about where it gets its sugar. Wine is made from grapes, beer can be made from grains such as wheat or barley (and often with hops flowers added), rum is made from sugarcane, hard cider is made from apples, and mead is made from honey (with water added). Corn, rice, and potatoes can also be fermented into alcohols.

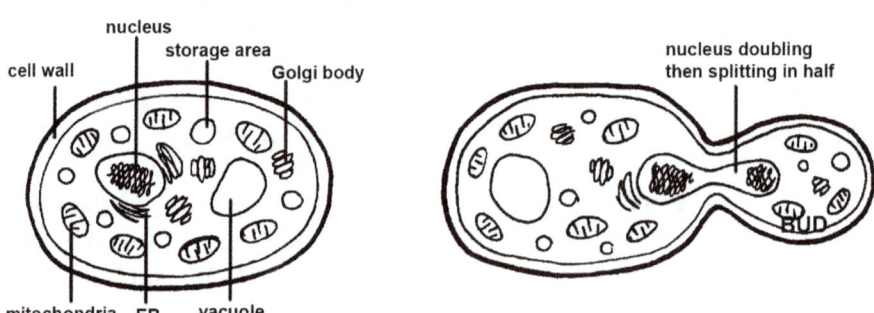

Yes, yeast is a type of fungus, but it is obviously not closely related to mushrooms. Yeast is a unicellular organism (made of single cells), whereas most other types of fungi are multicellular (made of many cells). Yeast cells are more complicated than bacteria cells, and have pretty much all the same cells parts as our human cells. In fact, sometimes research on yeast cells can shed light on things that go on in human cells. If scientists can use yeast cells instead of human cells, the research goes faster because yeast cells are so easy to work with. They can't file a lawsuit if the experiment ends up harming or killing them!

Yeast cells have all the standard cell parts: nucleus with DNA, Golgi bodies, mitochondria (energy makers), endoplasmic reticulum, vacuoles (empty "bubbles") and storage areas (similar to vacuoles).

If you have never studied cells, don't worry about this list. All you need to know is that they are different from bacteria.

Yeast cells don't come in male and female—they have no gender. They reproduce by growing a bud. As the bud grows, the yeast cell duplicates all of its cell's parts and puts some of them into the bud. The nucleus of the cell (containing the DNA) also is duplicated and an exact copy goes into the bud. When the bud reaches almost the same size as the original cell, it splits off and becomes an independent cell. The bud can actually start to form its own bud even before it reaches its full size. The budding process can be done in just an hour or two.

The lifespan of a yeast cell is determined by the number of times it can form a bud. When a bud finally separates from the "parent," it leaves a scar. Yes, even single-celled organisms can have scars. The scar is a place that has been damaged (though healed over) and will not be able to form a new bud. The parent yeast ends up with scars all over its body. When it is completely covered with scars and has no place left to form a new bud, it will stop dividing and die. When single-celled organisms die, they just dissolve and disappear.

Though they do not have genders, yeast can still do a very simple form of sexual reproduction where two cells will join together and combine their DNA. There is a survival advantage to mixing up their DNA. Normally, when conditions are good, yeast will just bud. But when conditions around them are not ideal (not enough food, too hot or cold, too dry, etc.) they will switch over to reproducing by trading their DNA. Small differences in DNA could possibly make some yeast cells better at surviving, and by sharing DNA, these advantages can be passed along to new generations.

This is an artist's sketch of a photograph of yeast cells. Those little rings on some of them are bud scars. Look at the cell that has the most scars It has a new bud growing on top, and the new bud is also starting to grow a new bud!

Now let's get out the Super Duper Magnifier and zoom in to the molecular level.

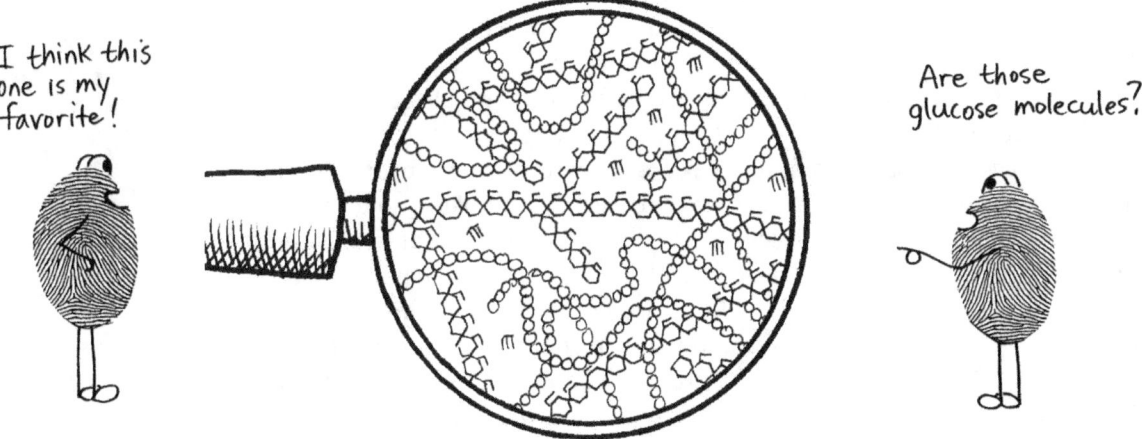

Our magnifier has simplified the molecules for us. We see a few tiny triglycerides here and there, but not a lot. We know there must also be water molecules scattered around, but our viewer has taken them out so they don't clutter the picture. Those strings of beads are actually protein chains made out of amino acids. You'll remember that very often amino acids are drawn as circles or balls. The lines of hexagons are glucose molecules all strung together. Let's start with these hexagons and do the proteins last.

We've already studied glucose. We saw how glucose and fructose can join to form sucrose. We also saw glucose bond to galactose to form lactose. Here we see only glucose molecules, but wow, are there a lot of them!

Glucose molecules are often shown as hexagons, with each vertex (corner) representing one of the atoms that make the ring. Glucose can also form a straight line, but the ring form is what is generally drawn. Sometimes a "flag" is drawn on top of the hexagon to represent that top carbon atom that is not part of the ring. This helps us keep track of "which end is up." Knowing which way the flag is pointing will be very important when we look at lettuce and spinach in the next chapter.

A simple way to draw glucose

Strings of glucose molecules are called **starch**. There are different types of starch, depending on whether the chains are just straight or whether they have branches coming off. The straight chains with no branches are called **amylose**. When the chain has branches, it is called **amylopectin** (am-ill-o-peck-tin). We can draw these in various ways, depending on how much detail we want to show.

We can keep it very simple and just show them as lines.	We can be more accurate and show all the "flags."	Or, we can decide to show the way they curl into helix shapes.
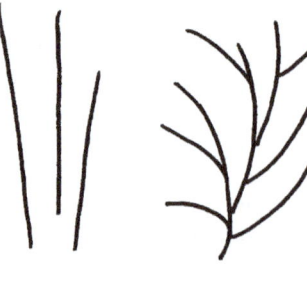	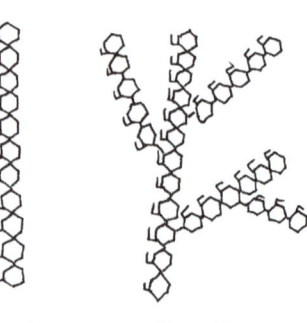	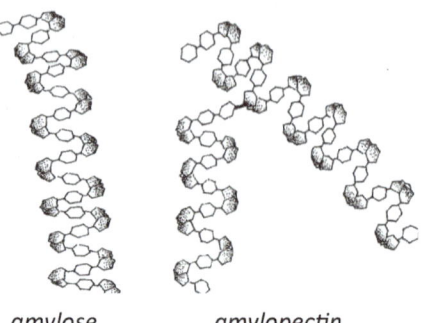
amylose *amylopectin*	*amylose* *amylopectin*	*amylose* *amylopectin*

Starch molecules are made by plant cells as a way to store food. First, the cells use the process of photosynthesis to turn carbon dioxide and water molecules into glucose molecules. Then the glucose molecules are assembled into long strings, and these strings are then put into storage "bags" called **starch granules**. Starch granules are sort of like a plant cell's pantry or storage cupboard. When the cell needs energy, it can take some starch out of storage and use enzymes to break apart the string, then split the glucose molecules, releasing energy. Plant cells can do glycolysis, but they can also tear apart the pyruvates if oxygen is available. Plant cells don't have to use any type of fermentation. Animal cells are the same way—they can deal with pyruvates if oxygen is available.

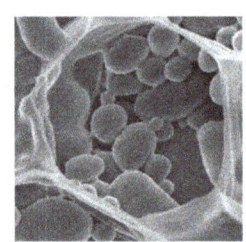
starch granules in a plant cell

The strings of starch in our bread came from wheat seeds. The wheat plant stored lots of energy in its seeds. When a seed falls to the ground and starts to grow, the baby plant will use some of the starch energy in the seed. However, a farmer got there first and harvested the wheat seeds. The seeds were ground into a fine powder that we call "flour."

Plants cells usually make more amylopectin than amylose. The ratio is typically 4 to 1. That means there are 4 amylopectins for every 1 amylose. The amount of amylose in a starch will cause it to behave in a certain way when it is cooked. Starch granules absorb water when they are boiled. They swell and grow larger and larger. Like an over-stretched balloon, they eventually burst. When they burst, the wet starch molecules come pouring out and start sticking together. This is why rice, pasta and potatoes can feel sticky. The temperature at which they burst will depend on what type of starch is being stored. Amylose is more resistant to boiling, so granules that have a lot of amylose will not burst as easily.

Rice is made of the seeds of the rice plant.

Rice is a good example of how starch ratios affect cooking. Long grain rice contains a lot of amylose, so when it is boiled, fewer granules break open. Less breaking means fewer loose starch molecules escape. Long grain rice tends to be fluffy and not stick together. Short grain rice, however, has very little amylose and a lot of amylopectin. When you cook short grain rice, it becomes so sticky that it will form balls very easily. This can be helpful if you are making sushi and you need the rice to form a solid layer in the sushi roll. Chefs can choose which type of rice they want to use in a recipe. If they want "not-too-fluffy-not-too-sticky," they can use a medium grain rice.

Yes, when you eat starch, your body breaks apart the molecules into individual glucose molecules. How fast this breaking down process happens is called the **glycemic index**. ("Glyc-" means "sugar.") Foods that are very high on the glycemic index are broken down into glucose very quickly. This would include white crackers, white potatoes, white rice, and white breads. Seeing a pattern there? Starches that are white tend to be easier to digest and therefore put more glucose into our systems in a shorter amount of time. Starches that are brown —the ones we call "whole grains"—take longer to digest. Whole grains include the brown layer that covers the seed, and this brown layer is not really digestible. Besides whole grains, other low glycemic starches include sweet potatoes, oatmeal, barley, and most vegetables. Why does this matter? As a general rule, it's best to keep the amount of glucose in our blood low and steady.

Starches are digested by enzymes. Amylose is taken apart by *amylase*. Amylase is found in both plants and animals. We can easily forget that from the plant's point of view, those starches are supposed to be used by the baby plant that will grow from the seed. Therefore, the plant must produce not only the starch but also the enzyme that will break it down. Amylase is found in seeds and fruits right there next to amylose, but it doesn't begin working until after the seed has fallen and is ready to begin growing. Part of the ripening process in fruit involves amylose breaking down starches.

Animal and human bodies make amylase, too. It's the main enzyme found in human saliva. We start digesting those starches even before we swallow them. You can test this by holding a cracker in your mouth without chewing. Not onlyl does it ges soggy and start to fall apart, but it begins to taste sweeter, too.

Amylase has the perfect shape for snapping apart the glucoses. The true shape of amylase is shown here on the right. It is made of amino acids but the artist has simplified the shape and not shown the amino "beads." Those curly places show that proteins, like starches, can form coils.

What would happen if we did not have an enzyme that could disassemble amylose? The long string would pass all the way through our digestive system, unchanged. In the next chapter we will meet a starch molecule we can't digest, though, oddly enough, it is still beneficial to eat lots of it.

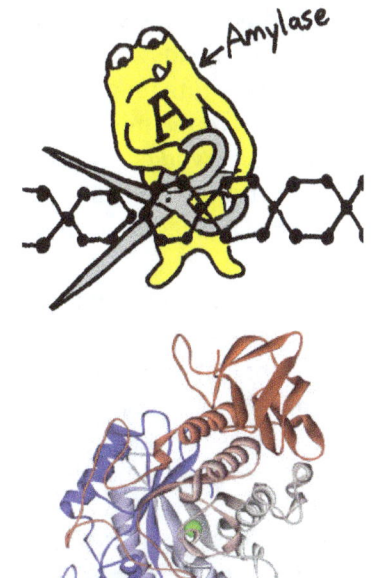

amylase

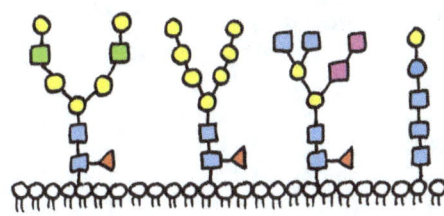

What will happen to the glucoses after they are snapped apart? Many of them will get used for energy, but not all of them. Glucose molecules are also used to build molecular structures. Strings of glucose molecules show up in a variety of places both inside and outside a cell. For example, they are part of the "post office" system inside a cell, being used in a way similar to address labels. Glycoproteins are used on the surface of a cell as identification markers.

49

Most people are surprised to find out that bread has a lot of protein in it. We tend to associate the word "protein" with foods like meat, fish, eggs, milk, and perhaps beans and nuts. Grains have quite a bit of protein in them, though. The protein is stored as long strings of amino acids, but we don't have a nice catch-all name for them like we do for glucose strings ("starch"). Storage proteins in plants can be called "storage proteins" but are usually known by their own names. For example, "avenin" is oat protein and "zein" is corn protein. We are going to look at two proteins found in wheat: **glutenin** *(GLUE-ten-in)* and **gliadin** *(GLIE-ah-din)*. Glutenin is usually shown as a long, straight molecule, and gliadin is usually shown as a curvy shape, or even sometimes as a circle.

You may be thinking that "glutenin" sounds a lot like a word we hear all the time now: "gluten." With all the hype about products being "gluten-free" you'd think that gluten was some kind of poison. Gluten is just the combination of the proteins glutenin and gliadin. It's actually gliadin, not glutenin, that causes problems for some people, but before we get into why gluten can be a problem, let's look at the positive side and find out what gluten does in bread.

If you've ever seen an expert pizza maker tossing a crust, you've seen what gluten can do. It's amazing to see that thin crust swirling around in the air, getting thinner and thinner yet not tearing apart. Gluten is stretchy and tough. Chefs can choose flours that have more or less gluten, and for pizza dough you definitely want a lot of gluten. Gluten is also what catches the carbon dioxide bubbles made by yeast, so it is a key factor in the fluffy texture of breads and rolls. Manufacturers of gluten-free bread must try to find a substitute for gluten. Often, sticky starches from potatoes or tapioca are added to the mix to increase the stretchy qualities of the dough. But if you've ever worked with gluten-free flour, you know that there really isn't a good substitute for gluten when it comes to making bread. Those gluten-free loaves of bread sold in stores are marvels of food engineering!

In dry flour, glutenin and gliadin are separate proteins. When water gets mixed into the flour, a major change happens. The water allows bonds to form between these two proteins. The amino acid responsible for this bonding is **cysteine**. Cysteine has, as part of its R group ("?"), a sulfur atom. (Shown on the next page.) Sulfur atoms like to bond with each other and form a **disulfide bond**. ("Di" means "two.") These disulfide bonds acts like bridges between the strands of protein. When a lot of glutenins bond to a lot of gliadins, we call the resulting substance **gluten**. The disulfide bonds are very strong so these long strands are stretchy and tough.

This illustration shows how glutenin and gliadin strands might interact. Glutenin is represented in green, and gliadin in blue. The red dots represent cysteine. This picture makes it easy to see how the cysteines form bonds that are like bridges. When you knead the bread dough, you are moving the protein strands around, and making more and more cysteines come into contact, so you get more bridges. More bridges means more stretchiness. Once the two proteins merge together, we stop naming each separate protein and just call the whole thing "gluten."

To understand why gluten can be harmful, we need to have a good understanding of how proteins work. You will remember that there are only 20 different kinds of amino acids. We briefly met a few of them in the last chapter, and we just met cysteine again on the previous page. Other amino acids you run into a lot include lysine, alanine, proline, valine, serine and glutamine. The amino acids are joined, end to end, to make a long string called a polypeptide. Glutenin and gliadin are polypeptides.

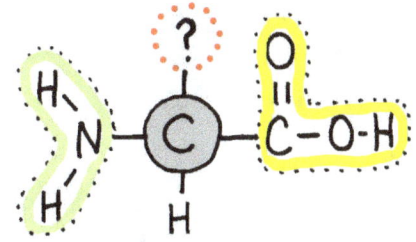

As we learned in the last chapter, each amino acid has parts that are the same: 1) a central carbon atom with an H attached, 2) a COOH (carboxyl) group, and 3) an NH_2 (amine) group. Every amino acid has these parts. What makes them different is the R group, represented here by a question mark. (Think of "R" as standing for the Rest of the molecule.) It is the R group that gives an amino acid its "personality." Some R groups have a small chains of carbon atoms and cause the amino acid to love fats and "hate" water, and therefore try to hide in the center of the protein molecule. Other R groups cause the amino acid to love water and be happy on the outside of a molecule. Some R groups are negatively charged and will be attracted to things that are positively charged. Some R groups are acidic or basic. A few amino acids have an extra amine (NH_2) group. Some have a benzene ring as part of their R group.

The interaction of all these various amino acid "personalities" will cause the protein to bend and twist and fold up into a specific shape. The number and order of the amino acids on the string determines the final shape of the protein. That shape will be useful somewhere in the organism that produced the protein, perhaps as an enzyme or a molecular shuttle bus, or as part of a muscle fiber or a strand of hair.

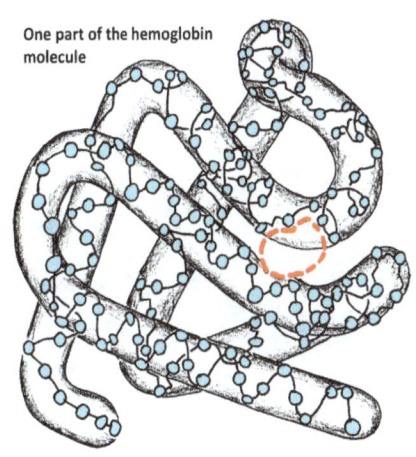

One part of the hemoglobin molecule

The protein shown here is part of the hemoglobin molecule that carries oxygen in your blood. You can see the chain of amino acids, though all the amino acids are colored blue instead of being different colors. The amino acids are arranging themselves, each one trying to get into a comfortable spot, and the result is this particular shape. The dashed red line shows where a "pocket" has formed. This pocket is where an iron-based molecule will sit. It will hold the oxygen atom.

Storage proteins like glutenin and gliadin are meant to be nothing more than a source of food. They must have an arrangement that **does NOT** fold up to be anything special. They must be random "nonsense" that is only good for being digested and recycled. They should **NOT** resemble any real proteins.

What would happen if a storage protein DID resemble a meaningful protein? This is the problem with gliadin. There is one place, at the end of the string, where the sequence of amino acids is too similar to a meaningful protein sequence.

We have white blood cells all through our body whose job it is to constantly be on the lookout for strange proteins that might be part of something that might hurt us—a toxin (poison), or a bacteria or virus. We call this army of white blood cells our *immune system*. Unfortunately, the end part of the gliadin molecule is recognized by some people's immune systems. The white blood cells don't have eyes or brains, so they can't see the sequence for what it is—just the end of a food protein molecule. To them it feels like an invader. This mistake happens in about 1 in every 100 people. The white blood cells begin a complicated attack process that ends up destroying innocent body tissues. When our

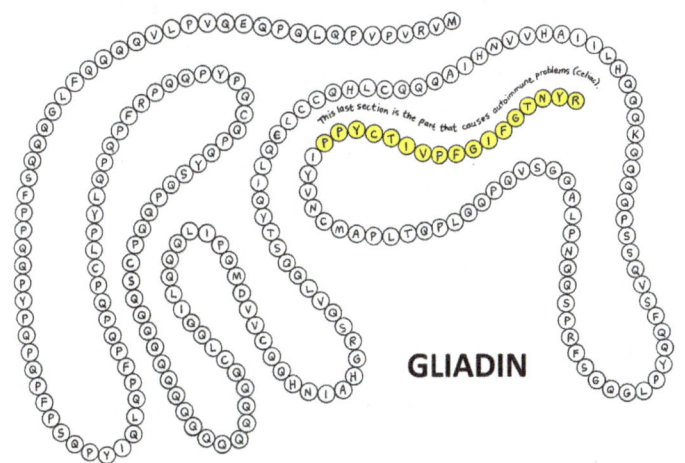

GLIADIN

Each letter represents an amino acid. Some letters make sense, such as P for proline, A for alanine, and C for cysteine. Some seem random, such as Q for glutamine. Look at how many Q's there are! Can you find the C's for cysteine? When two C's are close enough, they will try to bond. Where would cysteine bonds occur in this picture?

immune system attacks normal, healthy body cells this is called an *autoimmune disease*. The autoimmune reaction to gliadin is called **celiac** *(SEAL-ee-ack) disease*. People with celiac disease can be quite sick with not only intestinal problems, but also fatigue, headaches, muscle and nerve problems, and brain problems. The cure is to never again eat even one crumb of gluten.

Some people who don't have celiac disease also react badly to gluten. There isn't any name for this condition yet, so it is simply called "non-celiac gluten sensitivity." The most mysterious thing about gluten sensitivity is that before the 1980s, it was almost unknown. And now we have 7% of the population who can't eat gluten? What happened? No one knows for sure. Could it be human genetics? Plant genetics? Are plant scientists changing the sequence of the amino acids on the gliadin molecule? Some people blame a chemical named **glyphosate**, the active ingredient in the popular weed killer *Round-Up®*. Researchers are actively working to try to solve this mystery, and perhaps by the time you read this book, they will have found a good answer.

Interestingly, almost 23 percent of the population thinks they are sensitive to gluten but when they undergo very accurate testing, it turns out that they aren't. They might be reacting to something else in their food and just blaming it on gluten because we hear so much about it.

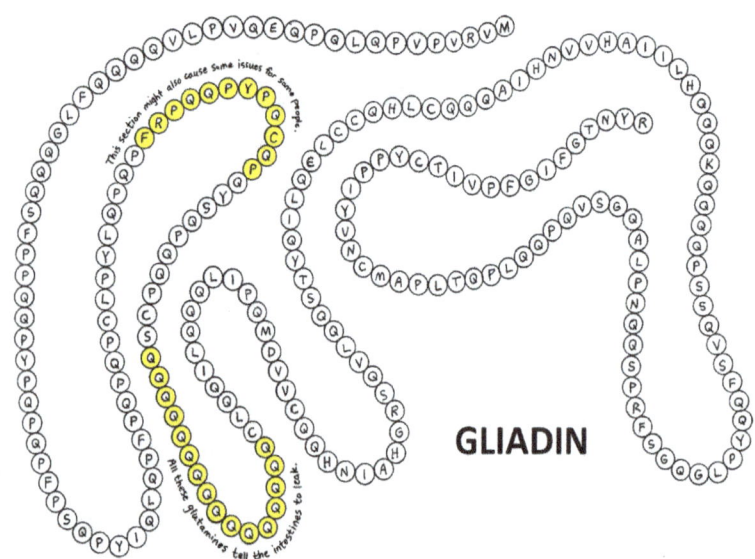

Gluten sensitivity involves a different place on the gliadin polypeptide, not the end that causes celiac. There are some places in the middle of the gliadin molecule that mimic another body protein, called **zonulin**. Zonulin is a messenger protein that tells the intestines to leak. Leaky intestines? This sounds like a very bad thing! However, there are times when you want your intestines to leak. As part of our immune defense system, white blood cells can be allowed to leak out of the blood and into the intestines to attack harmful bacteria. The cells of the intestines are very good at repairing the leaks, and most of the time things are quickly patched up. Unfortunately, 6 out of every 100 people have intestinal cells that aren't patching the leaks fast enough. Zonulin continues to be released and their intestines continue to leak, allowing molecules that normally stay in the intestines to get out into the blood. These molecules cause problems in various parts of the body.

There's one last thing we really must mention. Did you notice how golden brown the top of the roll was? We take "browning" for granted and don't realize that there is some complicated and interesting chemisty going on. This type of browning is known as the **Maillard reaction**. (Maillard is a French name, and is pronounced (*my-YAR*). The "d" on the end is silent. This name gets pronounced wrong very often, and it's easy to find bad examples on Internet videos.) The Maillard reaction is also responsible for the browning of meat. If you are a meat fancier, you might enjoy those cripsy browned edges where the meat was "seared" on the pan. The Maillard reaction can be quite delicious.

The Maillard reaction happens when sugar molecules get mixed up with amino acids and form bizarre "mutant" molecules. As the temperature rises in the oven, or in the pan, glucose molecules can open up and become a straight line instead of a circle, and amino acids can break apart. Amino acids, or parts of amino acids, start bonding to the sugars, making bizarre, nonsense molecules that are half-amino, half-sugar.

It would be like combining toys with appliances. Chop a bunch of toys in half, chop a bunch of appliances in half, then join them together randomly. What would we call these? They have no purpose. You can't call them toys anymore, and you can't call them appliances, either.

Real Maillard molecules don't look funny, like these crazy household combos do, but can still be interesting if you try to identify any remaining recognizable parts of sugars or aminos (such as R groups). Any of the 20 amino acids can be involved in a Maillard reaction, and the molecules can split and recombine in many different ways. The resulting molecules can then split and recombine yet again, so it's like a molecular "free for all" where mutant molecules zip about, combining and recombining to make hundreds of weird, unidentifiable molecules. The final result of this molecular chaos is... delicious smells and tastes!

Where might the sulfur atoms have come from in these Maillard molecules? And the nitrogens?

Crackers belong to a large category of baked goods called "quick breads." They are "quick" because in comparison to bread, they can be made very quickly. Yeast takes a long time to make bread rise. All those little single-celled yeast critters have to duplicate their DNA and their organelles and form buds that will become new yeast cells. This process can take several hours—which is pretty fast when you consider the complicated biological processes involved in cell division. But sometimes you need a method of making batter fluffy almost instantly.

Anything that will make a batter or dough fluffy (due to gas bubbles being created) is called a *leavening (lev-en-ing) agent*. Yeast is one of the slowest leavening agents. The fastest leavening agents are **baking powder** and **baking soda**. They are used in muffins, biscuits, cookies, cakes, pancakes, and some types of crackers.

Baking soda (also known as "bicarbonate of soda") is a white powder made of molecules with the chemical formula $NaHCO_3$. From the name, "bi-carbon-ate," we would expect to find two carbon atoms in the molecule, since "bi" means "two." But alas, the name was created using an outdated naming system and was never updated, so we are stuck with a name that does not seem to match the chemical formula. It's one of the rare oddities that chemistry students have to just accept and memorize.

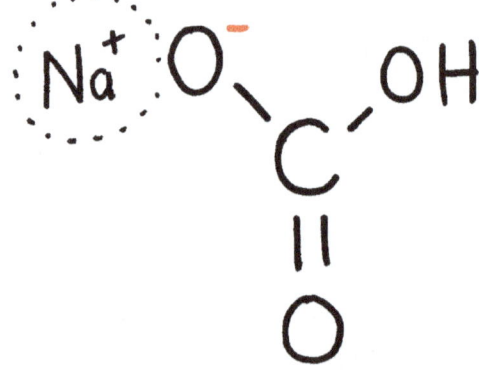

The (HCO_3^-) part of the molecule is called the **bicarbonate ion**, and it shows up quite frequently in the world of chemistry. As you can probably guess, this used to be a happy molecule with two hydrogens, H_2CO_3, then one of the hydrogens got fed up with never getting to have its electron, so it left the molecule and wandered off as nothing but a proton. The oxygen held on to H's electron, and there it is, in red, giving the oxygen a negative charge. Along comes a positively charged sodium ion (Na^+) and, as they say, "Unlike charges attract." (The sodium atom can't form the same type of bond that carbon does with oxygen. There isn't any sharing of electrons going on, just a strong electrical attraction.)

Where does $NaHCO_3$ come from? Though it might possibly be manufactured in a lab now, it was first discovered underground as a mineral rock, and is still mined today. The baking soda you buy in the store probably came from an underground mine. They send hot water down, dissolve the minerals, then pipe it up. A processing plant takes the water back out, and then puts the dry powder into boxes.

How does baking soda produce bubbles in batter or dough? You probably know the answer. Undoubtedly, you've seen what happens when you mix baking soda and vinegar—lots of bubbles form in just a few seconds. When baking soda is used in a recipe, you must also add an acid for it to react with. This is what happens when baking soda is combined with acetic acid (vinegar):

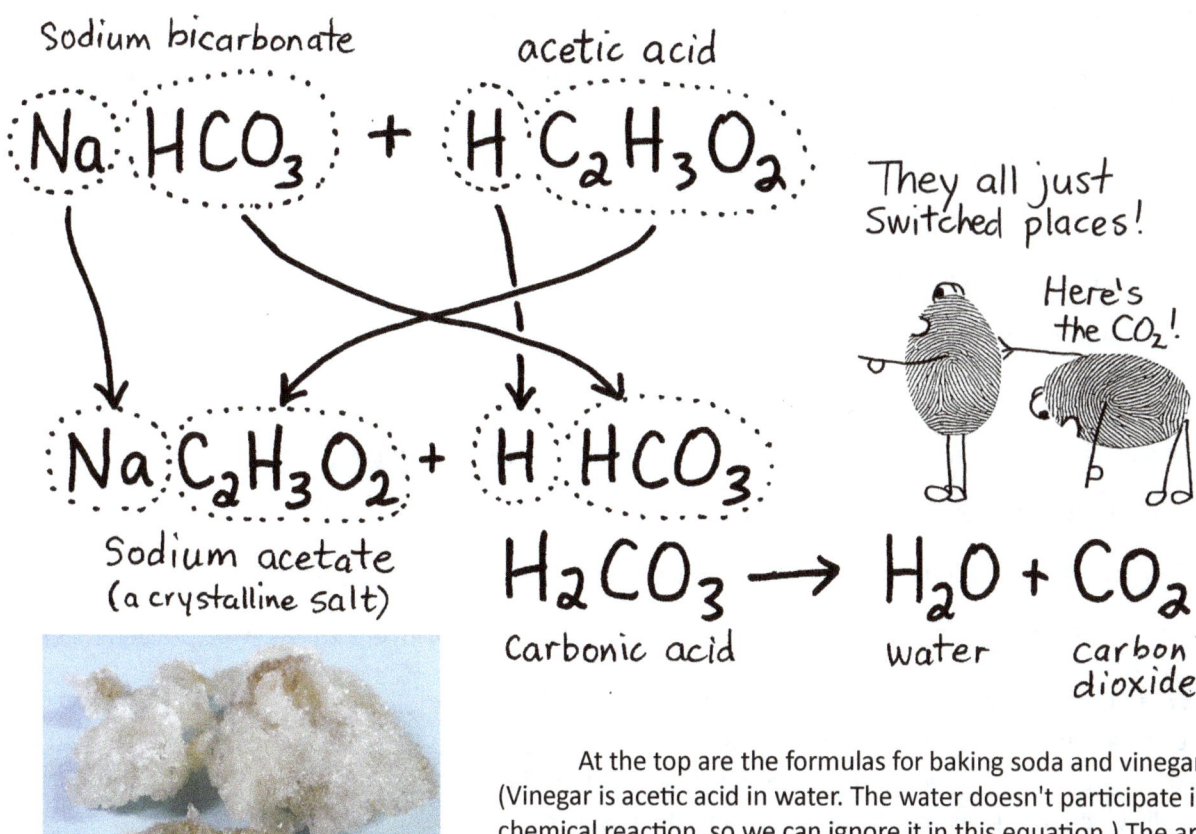

At the top are the formulas for baking soda and vinegar. (Vinegar is acetic acid in water. The water doesn't participate in the chemical reaction, so we can ignore it in this equation.) The arrows show you how the molecules rearrange themselves when given the chance. An H at the beginning of a molecule (as in $HC_2H_3O_2$) is likely to leave the molecule and wander off to join another molecule. Here, we see the H in acetic acid attaching itself to the bicarbonate ion, HCO_3^-, to form carbonic acid, H_2CO_3. Now we have two H's at the beginning of a molecule—double trouble! The carbonic acid molecule is very unstable and quickly falls apart to become H_2O and CO_2.

When we combine baking soda and vinegar, we get so excited about all the bubbles that we never think to look for anything else. As you can see from the equations above, a substance called **sodium acetate** is formed. (We already met this substance briefly on page 22 when we learned that when an acid and a base combine, water and a salt are produced.) Normally, the sodium acetate is dissolved in the water that is formed in the reaction, so we can't see it very well. However, if you boil down the leftover solution, taking all the water out, the sodium acetate will form beautiful crystals, as shown in the photograph. These are salt crystals, but not the kind we put in shakers. Food scientists do use sodium acetate in food products, however. "Salt and vinegar" flavored potato chips owe part of their flavor to sodium acetate. It is widely used (in small amounts) in many food products.

Baking powder is a combination of baking soda (sodium bicarbonate) and a powdered acid, such as **tartaric acid** (cream of tartar). When you add water to baking powder, the dry acid powder turns into a wet, highly active acidic solution which will begin to interact with the sodium bicarbonate. If you see the words "double-acting" on your can of baking powder, this means that the food chemists have designed the powder to be temperature sensitive. When you first add the water (or milk), you get a first round of CO_2 bubbles. Then, when you put the food into the oven, the heat will cause a second round of CO_2 bubbles to form while it is baking. Hopefully, the dough will become firm during the time when the reaction is at its peak, producing the maximum amount of bubbles. The trapped bubbles create a fluffy texture.

Comprehension self-check

1) The fats found in butter are mainly triglycerides. A triglyceride has three "tails" that are made of long chains of _____ atoms (with hydrogens attached).

2) The process of churning cream to make butter does what to the fat globules in it?

3) What comes out of the butter as the fats get packed together? _____ that has a tiny amount of protein and sugar floating in it.

4) What creates the acid found in "buttermilk"?

5) What does lactic acid do to milk?

6) What type of sugar is found in milk?

7) Why is butter yellow? It contains this molecule:

8) Name three foods that contain the molecule in #7.

9) Retinol is a form of vitamin ____, and is essential for the proper functioning of the _____ in your eye.

10) What is the most abundant protein found in milk?

11) This protein (in #10) floats around in milk in balls called _____ and the balls don't stick together because they all have a _____ electrical charge created by sugar/proteins called _____

12) When these protein balls are forced to clump together, this is called _____ milk. You can do this using a source of positive protons such as _____ or _____, or you can use an enzyme called _____ which comes from the stomach of a baby _____. This process also happens naturally when milk goes sour. What causes milk to go sour? _____ in the milk

13) When you make cheese, the watery stuff left over is called:

14) When "Little Miss Muffet sat on her tuffet, eating her curds and whey," what was she eating?

15) What microorganisms are used to make hard cheeses? _____ and/or _____

16) What do you call the process of splitting a glucose molecule, and what two molecules are the result?

17) Bacteria in milk produce _____ acid during the process called f_____.

18) The gas bubbles that make bread rise are made of _____ produced by _____ cells.

19) The cells in #18 also produce a waste product called _____, which disappears when bread is baked.

20) The cells in #18 are not bacteria. They are a type of:

21) Amylose and amylopectin are types of _____ and are made of _____ molecules.

22) The enzyme that can break apart amylose is called _____. Do plants make this enzyme?

23) Gluten is made of two proteins, called _____ and _____.

24) The sulfur-containing amino acid responsible for the creation of stretchy gluten is called:

25) People who are sensitive to gluten have immune systems that release too much _____, a messenger molecule that tells the intestines to become temporarily leaky.

26) This reaction is caused by broken pieces of sugars and amino acids bonding together: _____

27) If you use baking soda ($NaHCO_3$) in a recipe, what must you also add in order to get bubbles?

ACTIVITY 3.1 ALPHABET SOUP

Fill in a word for each letter of the alphabet. The clues below are in random order. Cross them out (or put a mark next to them) as you use them.

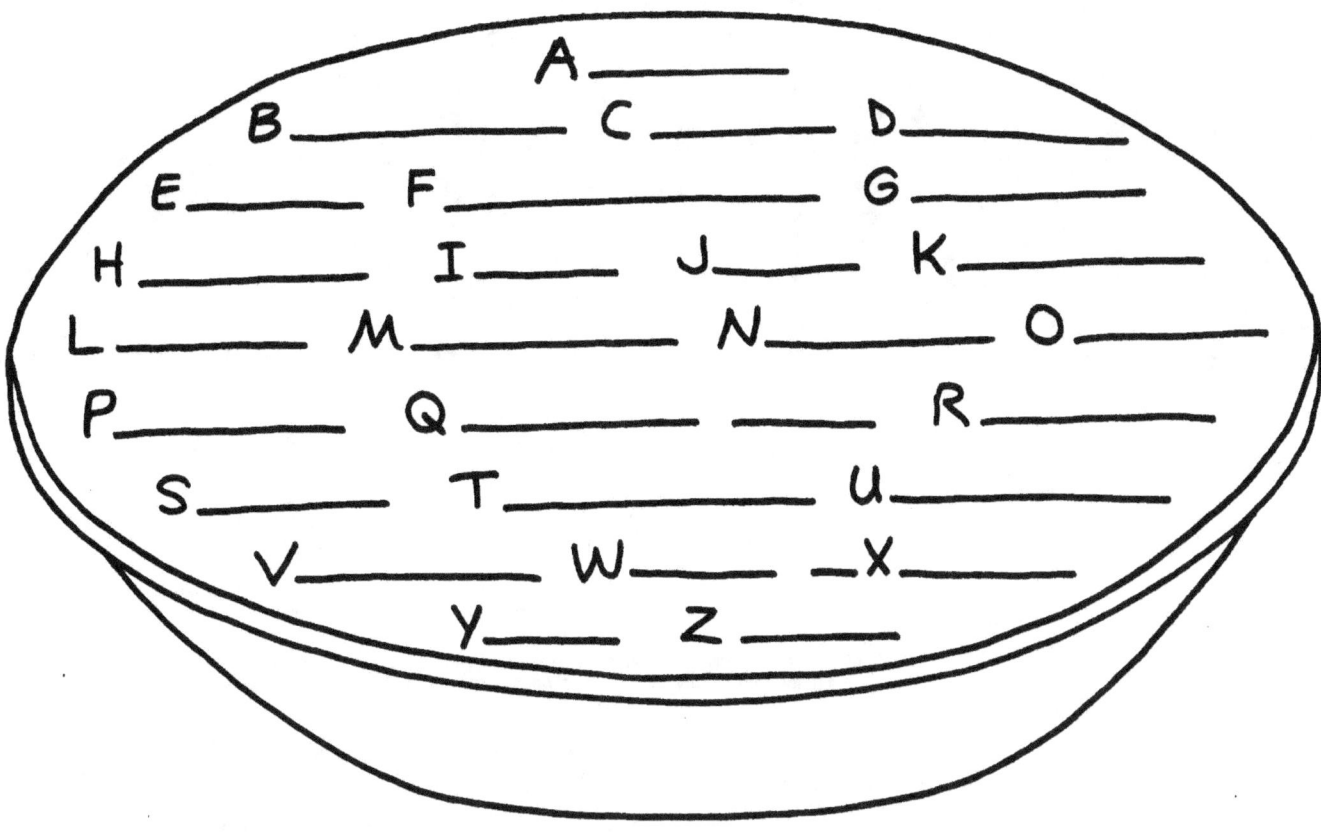

The tiny mineral clumps found in casein micelles are mostly made of calcium _____.
Beta-carotene makes food yellow or _____.
This amino acid is found in the strings hanging off casein micelles and it is very good at hanging on to sugars.
This molecule is created when enzymes add hydrogens to pyruvate.
Milk can be curdled instantly by using vinegar or lemon _____.
If you give an electron to a proton, you have made a _____.
Long strings of glucose molecule that can be broken apart by amylase
Oat protein
Corn protein
The watery liquid that is created during the cheese making process
The amino acid cysteine contains the element sulfur which allows proteins to form _____ bonds.
This is the room in your house where you do food chemistry experiments.
This organelle is found inside of animal and plant cells and harvests energy from pyruvate molecules.
An enzyme called Lactate Dehydrogenase puts H's onto pyruvate in order to _____ "shuttle bus" molecules.
Cornish _____ is a type of cheese.
Watery substance left over when butter is made.
This atom is used to patch the ends of the beta-carotene molecule when it is chopped in half.
The correct name for COOH
The part of gluten that causes problems
The casein micelles are surrounded by a _____ electrical charge.
Each of the 20 amino acids has a unique group of atoms located at the place where we put a _____ _____.
An empty bubble inside a yeast cell
The rate at which a starch can be broken down and absorbed by the body is called the glycemic _____.
A protein with a special shape that allows it to do a particular job
The molecule that fits into a protein holder in the cells in the retina of your eye
The process of turning pyruvate into lactic acid

ACTIVITY 3.2 STINKY CHEESE PUZZLE

Bacteria can be blamed for a lot of smells in the natural world—stinky feet, smelly arm pits, intestinal gas, bad breath, rotting meat, fermenting sauerkraut, and more. However, not all bacteria smell bad; a few actually smell flowery or fruity. But that's no fun. Smelly bacteria are more interesting. It's not just bacteria and molds that are used to make cheese—insects and arachnids are used, too. Yuck!

Learn the names of the stinkiest cheeses by using the key words at the bottom.

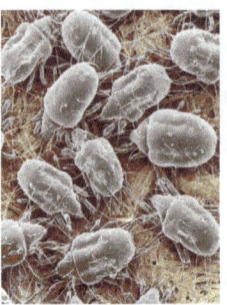

1) ___ ___ ___ ___ ___ ___ ___ ___ ___
   16 10   5   4   8 17 20 14 17

Made using *Brevibacterium linens*, a bacteria found on our feet, so it smells like stinky socks and armpits. Has been made for centuries in Belgium and Germany, and is probably the most famous stinky cheese.

2) ___ ___ ___ ___ ___ ___ ___ ___ ___
   18 21   5 14   5   4 14 17 13

This cheese is said to smell like a barnyard or like stinking laundry with hints of garlic. It uses raw, unpasteurized milk and is very runny because of the fluids produced by the fermentation process.

3) ___ ___ ___ ___ ___ ___ ___ ___
   14   1   2 10 22 22 14 22

Made with raw milk and rinsed in brandy. Internet rumors say it is so smelly that it was banned from public transport in France. So runny it has to be sold in boxes. Napoleon loved it.

4) ___ ___ ___ ___ ___ ___ ___ ___
   13 21 16 14 20 20 10   2

Made in Italy, smells like a combination of wet socks and wet grass. Is one of the oldest soft cheeses, dating back to the 10th century. Washed in seawater once a week. The taste is not so bad, a bit salty and fruity.

5) ___ ___ ___ ___ ___ ___ ___   ___ ___ ___ ___ ___
   22 13 10 11   6 10 11 20   4 10 22 15   2   1

Smells like wet hay, or the changing room of a football team. Based on a recipe that dates back to 1615. Uses milk from a rare breed of cow. The curing cheeses are washed in fermented juice made from "Stinking bishop" pears.

6) ___ ___ ___ ___ ___ ___ ___ ___
   17   2   7   8 14 12   2 17 13

Made using a blue *penicillium* mold that was orignally found in caves in southern France. Some people think it smells like rotten butter. Uses unpasteurized sheep milk so there is a small risk of *Listeria* food poisoning.

7) ___ ___ ___ ___ ___ ___ ___
   17 21 18 16 14 13 13 14

This is one of the runniest cheeses in the world. It is served warm, which makes is even runnier. It comes from the Alps of Switzerland where it is most often eaten at social gatherings. Some people think it smells like dirity feet.

8) ___ ___ ___ ___ ___ ___ ___ ___
    5 10   5   2 16 14 13 13 14

Made in France. It is formed into balls that look like cantaloupes. One of the key ingredients in its fermentation is the presence of mites, a tiny member of the spider family. The mites secrete chemicals that give it flavor.

9) ___ ___ ___ ___   ___ ___ ___ ___
   18 21 22   8   5 21 17   9   8

Made on the island of Sardinia, this cheese contains live maggots (larvae) from the cheese fly. The maggots can jump as far as 15 cm, so you have to hold your hand over the cheese as you eat. Or, you can put the cheese in a plastic bag an hour before you want to eat the cheese and the maggots will die.

10) ___ ___ ___ ___ ___ ___ ___ ___
     1 21 17   5 14 22 21 11

This Italian cheese is very smelly, but you have probably eaten some of it and enjoyed it. It is a very hard cheese, so hard that it must be shredded with a grater before eaten. It is often sprinkled over spaghetti.

KEY WORDS:

Amino acids are bonded together to make a ___ ___ ___ ___ ___ ___ ___
                                            1   2   3

The watery stuff leftover during the process of making butter: ___ ___ ___ ___ ___ ___ ___ ___ ___
                                                                    4   5   6

Whey is the ___ ___ ___ ___ ___ leftover when producing curds.
             7   8

This messenger molecule tells the intestines to leak, to allow immune cells to enter. ___ ___ ___ ___ ___ ___
                                                                                      9   10 11

Beta-carotene ___ ___ ___ ___ ___ ___ ___ orange and yellow light but absorbs green and blue light.
                     12   13

This alcohol is produced by yeast as they harvest energy from pyruvate molecules. ___ ___ ___ ___
                                                                                  14 15   16

When butter goes bad we say that is has gone ___ ___ ___ ___ ___
                                              17 18 19

White butter can be made from the milk of sheep and ___ ___ ___ ___
                                                     20 21 22

ACTIVITY 3.3   Third installment of "Chew It Over," a group game to be played during a meal

Here is another round of questions for you to use at a mealtime that you share with family or friends. These questions relate to the topics we learned about in this chapter. Again, you can use these questions in a variety of ways. You can be the quiz master and determine who gets which questions, or you can cut the questions out of the book and put them into a bag or bowl and let people choose a question randomly. The answers on are the back of this page.

CHAPTER 3: BUTTER, CHEESE, and BREAD

1) What makes the holes in Swiss cheese?

CHAPTER 3: BUTTER, CHEESE and BREAD

2) Which state in the U.S. is famous for its cheese making?

CHAPTER 3: BUTTER, CHEESE and BREAD

3) Queen Victoria received a giant wheel of Cheddar cheese as a wedding gift. About how much did it weigh?

a) 50 lbs (23 kg)   b) 500 lbs (225 kg)
c) 1,000 lbs (450 kg)   d) 10,000 lbs (4,535 kg)

CHAPTER 3: BUTTER, CHEESE and BREAD

4) How many varieties of cheese are there in the world?

a) 100   b) 2,000   c) 10,000   d) a million

CHAPTER 3: BUTTER, CHEESE and BREAD

5) What is the most popular (and most used) type of cheese in the world?

CHAPTER 3: BUTTER, CHEESE and BREAD

6) "Pule," the most expensive cheese in the world (300 US dollars per pound) comes from Serbia and is made from the milk of:

a) sheep   b) donkeys   c) elephants

CHAPTER 3: BUTTER, CHEESE and BREAD

7) What did ancient Egyptians use moldy bread for?

a) making soup   b) feeding cats
c) treating wounds   d) temple offerings

CHAPTER 3: BUTTER, CHEESE and BREAD

8) Where was a 2,000 year old loaf of bread found?

a) volcanic ruins in Italy
b) pyramid in Egypt
c) under a glacier in Norway

CHAPTER 3: BUTTER, CHEESE and BREAD

9) Many words are used to describe the smells of stinky cheeses. If you had to eat a stinky cheese, which of these would you choose?

funky, musty, goaty, tangy

CHAPTER 3: BUTTER, CHEESE and BREAD

STRANGE FACT: During World War II, when food was rationed, it was illegal to sell fresh bread in the UK. They thought the delicious smell of the fresh bread would cause people to eat too much all at once. The bread had to sit for 24 hours before it could be sold.

CHAPTER 3: BUTTER, CHEESE and BREAD

11) How many people do you know who don't eat gluten?

CCHAPTER 3: BUTTER, CHEESE and BREAD

12) Go around the table and try to guess each person's favorite starchy food. (anything made from rice, corn, wheat, or various types of flour)

1) Carbon dioxide bubbles from the bacteria that ferment the cheese as it hardens.
2) Wisconsin
3) 1,000 lbs (450 kg)
4) 2,000
5) Mozzarella
6) donkeys
7) treating wounds
8) volcanic ruins in Italy

4: Salad

Wow! Where do we begin? You might be about to reach for one of those bottles of salad dressing. How about if we start with those? It looks like we have an Italian style dressing and a French dressing. They look very different (the Italian is fairly clear, the French is orange and opaque) but their basic chemistry is similar.

All salad dressings are a type of solution called an **emulsion**. An emulsion is made of a polar liquid and a non-polar liquid held together by an **emulsifier**. Good thing we already know what a polar molecule is! In chapter one we learned why water is a polar: the electrons are shared unevenly. The poor hydrogens hardly ever get to see their electron because oxygen is such a bully. There are many other polar molecules, beside water. A simple way to test for polarity is to see if the substance will dissolve in water. If it does, it's polar. There are so many substances that dissolve in water, that water is often called "the universal solvent."

Non-polar substances won't mix with water. You can probably name some: cooking oil, the fat found in meat, butter, shortening, petroleum jelly, motor oil, etc. Anything greasy or oily is non-polar. At the molecular level, a non-polar substance usually contains strings of carbon atoms, like we saw on our triglycerides. The carbon and hydrogen atoms in a fat molecule are not at all interested in interacting with water molecules. In fact, we call them "hydrophobic." The word "hydro" is Greek for "water," and "phobic" is Greek for "fear or hate." Sometimes we say that non-polar substances "hate water." Molecules have no feelings, of course, so they can't love or hate anything, but it does make the chemistry more memorable.

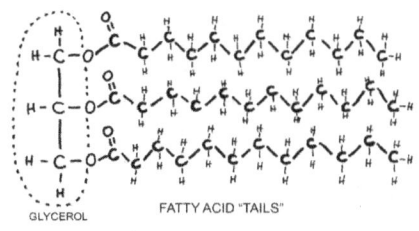

GLYCEROL   FATTY ACID "TAILS"

HYDROPHOBIC FAT   WATER MOLECULES

A salad dressing is an almost equal amount of a non-polar substance (like olive oil) mixed with a polar substance (like vinegar, lemon juice or water). How will this work? We are mixing two substances that "hate" each other. If you've never done this, try it at home. Mix some oil and water or vinegar, shake the bottle, then see what happens. The oil will float to the top and stay there, and the water or vinegar will happily rest on the bottom. They will stay like this until you shake the bottle again. Though salad dressings do need to be shaken, they stay mixed a little better than this. So, what's the secret ingredient?

The secret ingredient is an *emulsifier*. Common emulsifiers found in dressings are egg yolks, mayonnaise, mustard, and honey. These substances just happen to have the right chemistry.

Emulsifier molecules have a part that is polar and a part that is non-polar. This means they can interact with both water <u>and</u> oils (polar and non-polar). When enough emulsifier is mixed into the dressing, the oil and water don't separate nearly as much. Or at least not for a while. If you've ever gotten an old bottle of dressing out of the fridge, one that has been sitting for months, you may have noticed that the ingredients have settled out into layers.

The last essential ingredient in a dressing is something (or several things) to give it a particular flavor. Our Italian dressing probably has bits of dried herbs such as oregano, basil, thyme or rosemary, and garlic. Many dressings also have salt, pepper or sugar. Dressings that are thick and white may also use cream or sour cream.

With these four ingredients (oil, water or vinegar, an emulsifier, and spices) you can make your own dressing. The dressings you buy at the store often have preservative chemicals added, too, so that they won't go bad if they have to sit on the store shelf for many weeks or months. If you read the list of ingredients you can find out exactly what preservatives they've used, but you might be a bit disappointed to find that they don't list all the herbs and spices. It might just say "spices." They are allowed to keep some parts of their recipe a secret.

Sure, no problem. Mayonnaise is probably the most well known and widely used emulsion in many parts of the world. Emulsions don't have to be liquid like salad dressings; they can be soft solids like your mayonnaise. You can make your own mayonnaise from scratch using any type of vegetable oil, eggs, lemon juice or vinegar, and mustard as your emulsifier.

Of course, not everyone can eat eggs and mustard. Food scientists have discovered substitute emulsifiers that almost everyone can eat. For example, here is a widely used emulsifier that has a name that might look scary when you read it on a label: "Monoglycerides and diglycerides." Are they dangerous?

Jog your memory a bit and you will hopefully remember seeing a similar word: triglyceride. That's our jellyfish-looking fat molecule (shown as a funny cartoon at the top of this page). "Di" means "two," and "mono" means "one." If you chop one leg off a triglyceride, you get a diglyceride. Chop off another leg and you get a monoglyceride. The place where the leg came off can grab on to something else, often a polar molecule. The membranes around all your cells (and the membranes around the fat globules in milk) are made of diglycerides that are holding onto a polar phosphate molecule.

Monoglycerides and diglycerides are used in many food products including ice cream, chewing gum, bakery products, whipped toppings, and nut butters. They are particularly helpful in industrial bread baking, as they improve the volume and texture of the loaves.

Enzymes make monoglycerides and diglycerides

We are going to take a short break from chemistry and do some botany (the study of plants). Salad usually contains a number of plant parts: leaves, stems, roots, buds, fruits and seeds. You can probably name most of the parts in your salad. Lettuce and spinach are leaves, celery is a stem, carrot is a root, and broccoli crowns are buds. However, you might not identify the fruits: peppers, cucumbers and tomatoes. Scientists and chefs have different definitions of the words "fruit" and "vegetable."

To a scientist, anything with seeds is a fruit. Peppers, cucumbers and tomatoes all contain seeds, so a botanist would say they are fruits. The scientific definition for vegetable is the "vegetative" parts of a plant, meaning those parts that are <u>not</u> involved in reproduction—leaves, stems, and roots.

To a chef, the word "fruit" is restricted to the scientific fruits that are also sweet, and "vegetables" are non-fruits, no matter what part of the plant they come from. A chef's vegetables include roots, leaves, stems, fruits, and seeds. (Peas and corn are seeds.)

We prefer to eat plants at various stages of their life and can easily forget that they all have the same structures. For example, we think of lettuce only in its leaf form, but lettuce plants produce flowers, fruits and seeds. In fact, hundreds of years ago, lettuce was used for its seeds, not its leaves; oil was pressed out of the seeds. It was only later

lettuce seed

that people decided to start eating the leaves. We are very used to seeing heads of broccoli with all those tiny green buds, but if you have ever grown broccoli in your garden, you will probably have seen some heads that were not picked early enough and started to bloom with tiny yellow flowers. Bean sprouts are tiny plants that we eat whole—stems, roots and leaves. In some cases, there is a good reason why we eat only one part of a plant. Tomato and potato stems and leaves are mildly toxic.

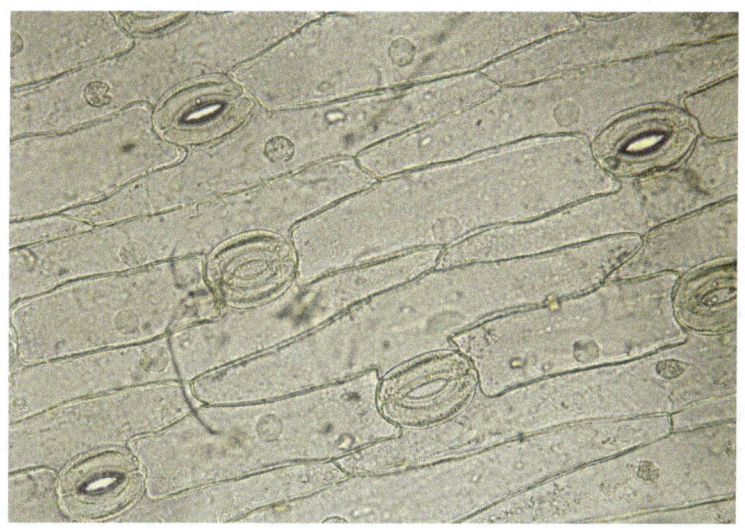

What do your lettuce and spinach leaves look like under the microscope? It depends on how we look at them. If we look straight down on a leaf, we will see a flat, green surface made of many long rectangles. The rectangles are the plant cells. If we look at the underside of a leaf, we will see the same rectangles, but we will also see tiny holes called **stomata**. The holes are made of two **guard cells** that can open and close the hole, controlling the flow of air in and out of the leaf. They often close during the day when it is hot and dry, and open at night when it is cooler.

If we cut the leaf and look at the thin side we just cut, we will be able to see the leaf's internal structure. There are different types of cells on the top, middle and bottom.

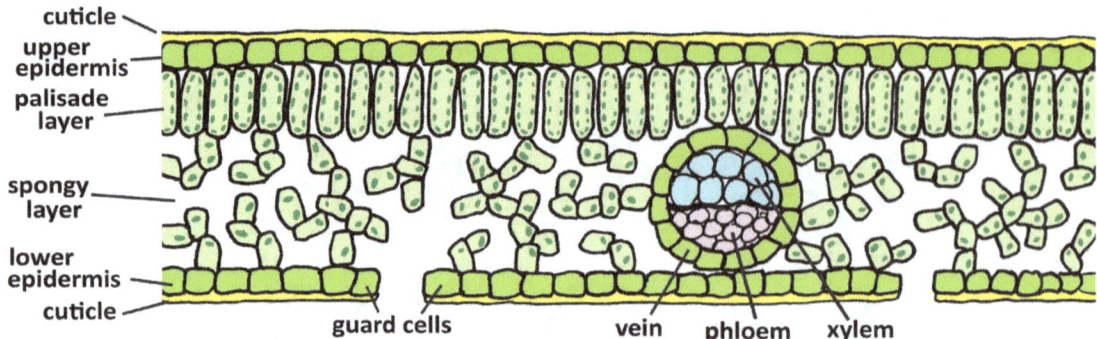

The yellow cuticle *(kew-tick-uhl)* layers are made of a waxy, waterproof substance. They keep the leaf from drying out or getting too wet. The upper and lower epidermis layers form the large flat surfaces of the leaves. The palisade layer is the primary site for photosynthesis. Those tiny green dots are chloroplasts. It is inside the chloroplasts that sunlight, carbon dioxide and water are turned into glucose sugar molecules. The spongy layer has lots of air space. (The cells are not really floating—cross sections just look strange because we don't see those cells in a 3-dimensional context.) The spongy cells have a few chloroplasts, but not as many as the cells in the palisade layer. The circle is a vein that has been cut. The inside of the vein has two types of cells. Xylem *(zie-lem)* cells transport water, and phloem *(flow-em)* cells transport sugary sap along with some minerals. This is one of the smaller veins in the leaf. You would need a microscope to see it.

The drawing above does not show all the organelles inside the cells. Let's take a close-up look at one palisade cell and compare it to a yeast cell.

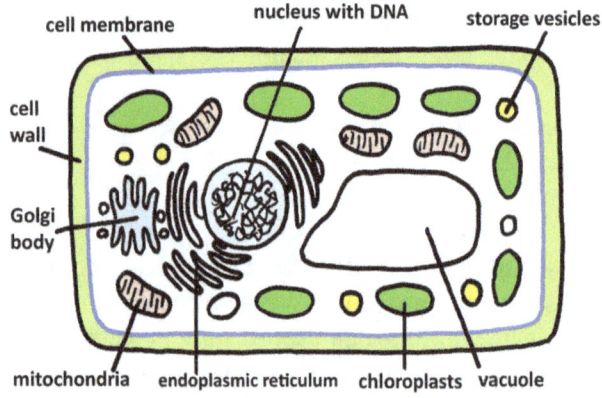 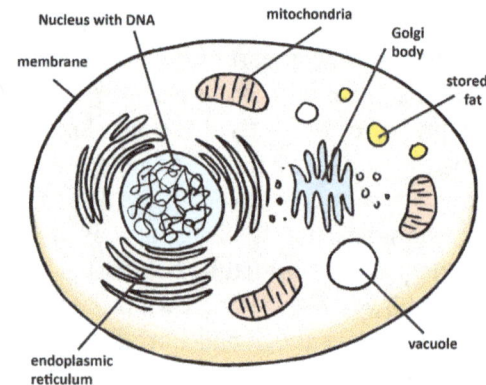

We don't have time to go into great detail about what all these organelles do. (If you want to know all the details, check out the book titled "Cells" by the same author.) Most of a cell's parts are made of things we've already learned about: proteins (strings of amino acids), fats, and sugars (usually in long strings). We'll learn more about the cell wall and the chloroplasts in a minute.

The fourth major ingredient is the cell's nucleic acids: DNA and RNA. The cell's DNA stays inside the nucleus, and RNA can be found both inside and outside the nucleus. DNA and RNA are made of three parts: a sugar, a phosphate and something called a base. There are only 5 different bases, and these bases are the same for all living organisms. The only difference between your DNA and a plant's DNA is how the bases are arranged. So as your body digests the plant cells in your salad, some of the lettuce and spinach DNA rungs could end up as part of the DNA in one of your own cells!

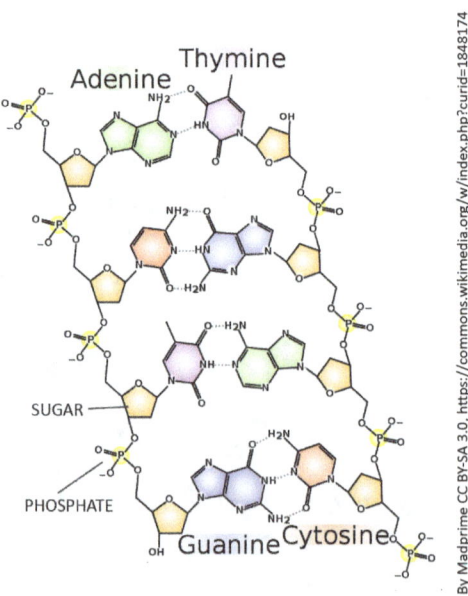

If we look inside a chloroplast, we find hundreds, or perhaps thousands, of things that look like stacks of green pancakes. The outside of these flat **thylakoid** discs is made of the same stuff as the cell's thin membrane just under the thick cell wall. (The yeast cell also has a membrane around it, but no cell wall. Only plant cells have walls.) Cell membranes are made of a double layer of molecules called **phospholipids**, which look a lot like our two-legged diglycerides. These "legs," which are properly called "tails," are made of fatty acids—chains of carbon atoms with hydrogens attached. The round part (shown as yellow circles in the diagram below) isn't really round, but represents an area where the third glycerol "clip" is holding on to a water-loving phosphate molecule.

The purpose of these thylakoid membranes is to be a platform into which a lot of amazing little nano-machines can be embedded. Some of these "machines" function as pumps or shuttle buses, and one of them, called ATP synthase is a proton-driven motor that recharges little energy molecules. Most of these machines are made of protein (strings of amino acids). The DNA holds the code for how to string amino acids together so that the string will fold up into a useful gizmo that will perform a task.

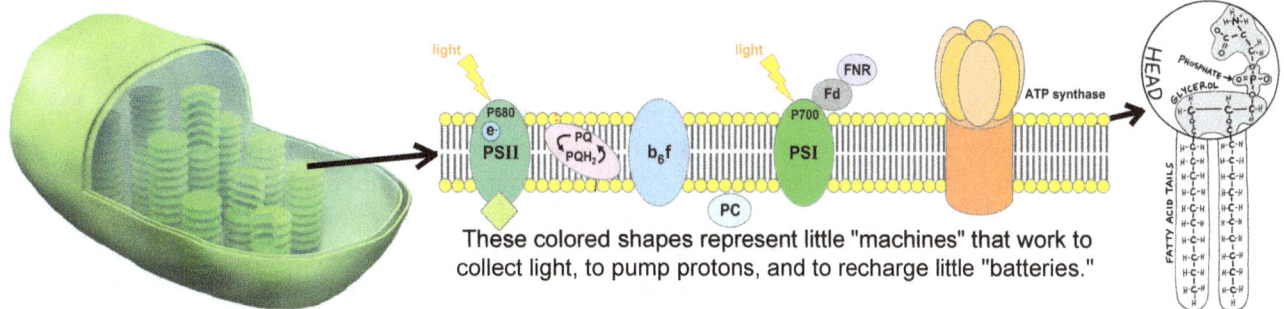

These colored shapes represent little "machines" that work to collect light, to pump protons, and to recharge little "batteries."

The green ovals represent a complicated light-collection system. One of the key molecules in these systems is **chlorophyll**. Since chlorophyll reflects green light, leaves appear green. These "photosystems" contain a lesser amount of other light-collecting molecules such as **beta-carotene** (orange), **xanthophyll** (yellow) and **anthocyanin** (red). The green chlorophyll is so strong that it often masks over these other color molecules. If the chlorophyll is removed, you can then see these other colored pigment molecules. This is what happens in the autumn when green leaves turn red, orange or yellow.

CHLOROPHYLL

BETA-CAROTENE

Because this is a food chemistry book, we are not going to take the time to learn how these fascinating cellular machines work. Instead, we are going to break them and tear them apart. When you chew, some of the plant cells will break open and the organelles will spill out. Digestive enzymes will then go to work and reduce them to piles of fats, amino acids, sugars and nucleic acids. (Yes, it does seem a shame to destroy all these fascinating little machines, but plants are always busy making new ones.) Amino acids are the same in all organisms, so your cells can use the digested amino acids taken from the plant cells to build their own cellular machines.

organelles spilling out of cells

As we've already seen, **beta-carotene** molecules can be chopped in half and used in your eyes. The square-ish part of the chlorophyll molecule can be re-purposed by taking the magnesium (Mg) atom out of the center and replacing it with an iron atom, making it into **heme** *(heem)*, the molecule in your red blood cells that carries oxygen. Other pigment molecules are used by our cells as natural sunscreens, or as chemicals that help our immune system to be more efficient.

Beta-carotene, chlorophyll and other pigments are part of very large group of molecules called **phytonutrients** *(fie-toe-nutrients)*. Phytonutrients are the fifth category in our list of types of nutrients found in plants. For the most part, phytonutrients are not proteins (not made of amino acids), and they are not classified as fats or sugars, though they might have some chains of carbon atoms (like a fatty acid) or rings of carbon (like glucose). They form a very large and diverse category, with some scientists claiming there are as many as 25,000 of them.

"Phyto" is Greek for "plant."

Phytonutrients, also known as **phytochemicals**, are classified by their molecular shape and by what types of atoms they contain. Some are long and straight, like beta-carotene. Others have many hexagons or pentagons. Molecules containing sulfur have their own group. We'll look at just a few general categories. (If you want a lengthy list, check out the Wikipedia article titled "List of phytochemicals in food.")

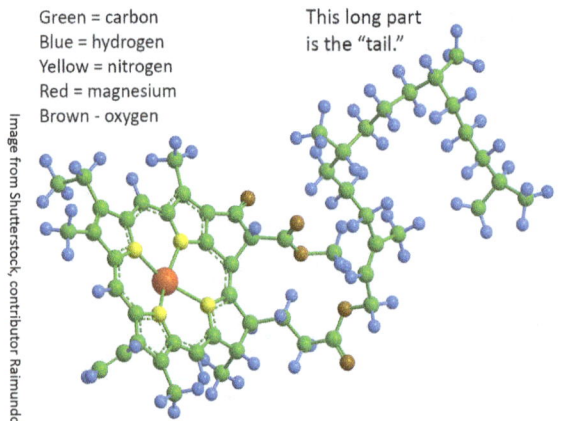

Green = carbon
Blue = hydrogen
Yellow = nitrogen
Red = magnesium
Brown - oxygen

This long part is the "tail."

The **chlorophylls** have their own category. There's chlorophyll-A, chlorophyll-B, and others. They are all some shade of green, and their name comes from the Greek word for green: "chloro." ("Phyll" means "leaf.") Chlorophyll molecules have a square-ish part that has an "X" in the center, made of four nitrogen atoms and one magnesium atom. ("X marks the spot" where light energy is first captured for the process of photosynthesis.) There is also a long "tail" made of carbon and hydrogen atoms. Does it remind you of a fatty acid? The tail is fat-friendly and helps to hold the molecule in place in the membrane.

Beta-carotene belongs to a group of phytonutrients called **carotenoids** *(care-OT-in-oidz)*. All of these molecules are orange, yellow or red. Many of them have complicated names that we'll never remember, but here are two names that are fairly well-known: lycopene and lutein. These molecules are often included in multi-vitamin pills. They are said to have anti-cancer properties as well as positive effects on the eyes and the immune system. The molecules in this group tend to be long and straight (like a carrot) and are often at least somewhat symmetric (the same on each side).

BETA-CAROTENE

LYCOPENE

Carotenoids tend to be long and symmetric.

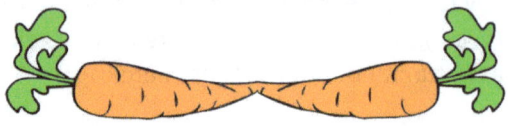

Polyphenols *(polly-FEE-nols)* form the largest category of phytonutrients. Subcategories include flavonoids, tannins, curcuminoids, and resveratrol. These names may sound difficult and strange, but in the world of food chemistry they are words that everyone knows. The word root "poly" means "many" and "phenol" is a hexagonal "ring" of carbon atoms with an -OH hanging off. So all the molecules in the polyphenol group have more than one of these carbon rings. Some of them have dozens of these rings.

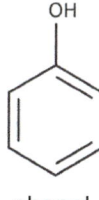
phenol

Flavonoids are found in many fruits and vegetables, and also in coffee, tea, and chocolate. ***Anthocyanin*** flavonoids are red, blue and purple and give those colors to the plants that contain them. Studies find that, on average, people who eat a lot of flavonoids experience better overall health than those who do not. Flavonoids help to slow the aging process in a number of ways, including reducing inflammation and helping to catch "free radicals" (dangerous broken molecules that can damage healthy body molecules).

Curcumin *(ker-KOO-min)* is a chemical found in ***turmeric***, a bright yellow spice found in many Indian recipes. Curcumin reduces inflammation, stops cancer cells from growing, and lowers the amount of cholesterol (a type of fat) in the blood.

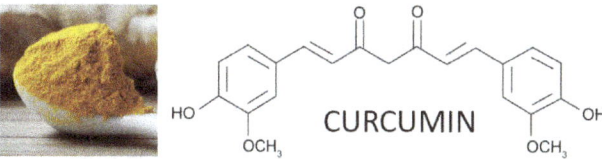

Resveratrol *(res-VARE-i-trol)* is most abundant in dark red or blue fruits such as grapes, blueberries and cranberries. It seems to have anti-inflammatory properties as well as the ability to help your body get rid of dangerous "free radical" molecules that have unpaired oxygen atoms. Oxygens that are either on their own or are left hanging on the edge of a broken molecule can wreak havoc on other molecules in their neighborhood. Substances that can capture these dangerous oxygen atoms are called ***antioxidants***.

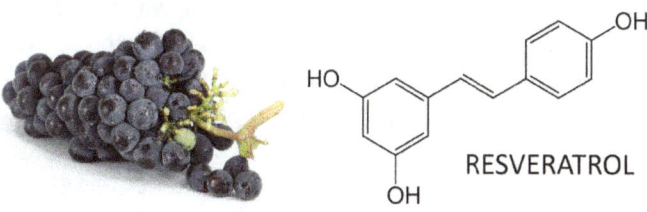

Tannins get their name from the process of tanning leather. For many centuries, the bark of oak trees was the source of important chemicals for the tanning process, though no one really knew what they were until chemists in the 1800s were able to extract the various tannin chemicals, isolate them, and give them names.

TEA LEAVES

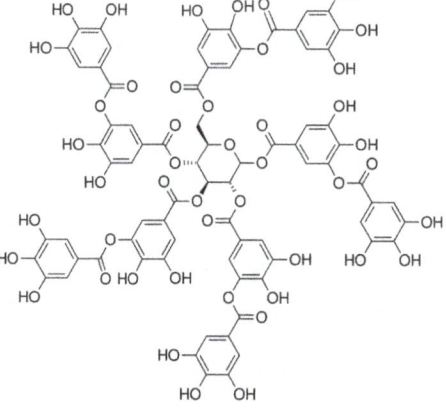
TANNIC ACID

Tannins are found in great abundance in tea leaves, and are responsible for tea's "tangy" taste. Tannic acids are made by plants as pesticides—chemicals that are toxic or distasteful to organisms that want to eat the plants, such as insects or bacteria. Tannins, as well as many other phytochemicals, are part of a plant's defensive strategy to avoid being eaten.

Polyphenols interact with your cell membranes in ways that increase their strength and their connection to other cells.

Many vitamins are similar to some phytonutrients. We've already seen the connection between vitamin A (retinol) and beta-carotene. Enzymes in our bodies can modify beta-carotene and turn it into vitamin A. Vitamins are sort of like phytonutrients we can't live without. We can survive without ingesting all the different types of polyphenols and tannins. But if we try to live without vitamins, we will eventually develop a deficiency disease. Diseases due to lack of vitamins are actually what led to their discovery. The story of vitamins starts out in the so-called "Age of Discovery," when European explorers began to make long voyages by ship.

Starting in the 1500s, Europeans set out to explore the world in wooden sailing ships. Often, their real mission was to discover where certain spices were grown, such as pepper, cloves, and cinnamon. These spices are so common now that is hard to imagine a world in which a teaspoon of cloves was worth hundreds of dollars. We shake pepper onto our food without a thought, not knowing that in past centuries battles were fought over this spice.

The voyages were often very long and the sailors spent months at sea. Fresh food only lasted a week or two and after that it was nothing but hard tack biscuits made of wheat flour and possibly some salted preserved meat. After several months at sea, the sailors would begin to get sick with a disease called *scurvy*. Their gums would start to bleed, then they would get sores all over their body that would not heal. Eventually the sores would become infected and they would die. (At the time, they noted that it was rarely the officers on board that came down with scurvy, just the sailors. Hindsight allows us to solve this mystery. Only the officers were allowed to drink wine, which is made from grapes, a source of vitamin C.)

In 1747 a Scottish surgeon named James Lind put forth his theory about how to prevent scurvy. He recommended that citrus fruits such as lemons and limes be added to the food provisions. These fruits did not go bad quickly and could even be sliced and dried. Dr. Lind did not know about vitamin C, only that citrus fruits seemed to prevent scurvy. The British navy took his advice and the incidence of scurvy dropped dramatically.

Not everyone was convinced, however. Some very influential people held that good hygiene (staying clean), regular exercise and maintaining a positive attitude were the keys to keeping sailors healthy.

Other people thought scurvy might be caused by spoiled food. These ideas were put to the test in the 1800s when a number of Arctic expeditions decided to sail without bringing any citrus fruits along. The results were disastrous.

In 1881, a Russian scientist, Nickolai Lunin, did one of the first experiments to test the idea that disease could be caused by a lack of nutrients. (The reason that this idea was not being explored by more scientists was that germs had been discovered during this time period, and scientists had jumped to the conclusion that they now knew the root cause of ALL disease.) Dr. Lunin set up an experiment with mice where he controlled their diet, allowing some baby mice to drink milk, but feeding other baby mice a concoction he'd mixed up that had fats, sugars, and amino acids—all the nutrients they required, as far as he knew. Of course, the baby mice who drank this substitute milk died. Dr. Lunin concluded that milk contained other ingredients essential to life, besides fats, sugars and proteins—substances that had not yet been discovered. Many scientists disagreed with his conclusion, and thought the mice had died for another reason. It is important to remember that scientists don't always agree, and can often be resistant to new ideas.

Aren't these guys adorable? (Even if you don't like rodents, at least it's a break from chemistry!)

After this, other researchers began to design similar experiments, and finally the results were overwhelmingly in favor of the theory that foods contain substances other than protein, fats and sugars, that are essential to life.

In the early 1900s, many experiments were performed on small animals to try to identify the "anti-scurvy" substance. The breakthrough came when experiments were done on guinea pigs instead of rats or mice. Guinea pigs and humans are the only mammals whose cells are unable to produce vitamin C—they must get it from their diet. Scientists were able to perform experiments on guinea pigs that they could never do on humans. Finally, in 1932, the discovery of the "anti-scurvy" substance was announced. It was named **ascorbic acid**, with "a" meaning "against" and "scorbic" coming from the Latin term for scurvy: "scorbuticus." Only later was it called vitamin C.

The author of this book used to raise long-hair guinea pigs, like this one. They got plenty of foods with vit. C.

Meanwhile, research was also being done on other diseases that seemed to be caused by a lack of nutrients. During the late 1800s, a process for refining and polishing rice was invented. The brown hull around the rice was removed, leaving a beautiful white grain with a mild flavor. People in Eastern Asia gave up eating natural brown rice and began eating nothing but white rice. The Japanese Navy was especially hard hit, with many sailors dying of a condition they called "beriberi," which caused the heart, brain and nerves to malfunction. Just like in the British Navy, it was the low-ranking sailors who were most affected, not the officers. The officers were fed meals with a greater variety of foods, but at the time no one thought this was important.

In 1897, a Dutch physician used experiments on chickens to show the link between polished rice and beriberi, though at the time he was unable to determine the essential substance in the brown rice hulls. Eventually it was shown to be a chemical called thiamine, which later was named vitamin B1.

Chickens were used in the research that discovered thiamine, B1.

In the early 1900s, there was a race among researchers to be the first person to discover a new essential nutrient. Communication at this time was very slow. The telegraph was the only way to get news to other parts of the world quickly, but it was limited to short sentences, and wasn't a method of communication that worked well for scientists who wanted to explain their research. It was not uncommon for scientists to be completely unaware of research that people in other parts of the world were doing. Two researchers could make the same discovery simultaneously and not know it. This led to many controversies and arguments among the scientists of the world about who should be allowed to claim credit for the discoveries and possibly earn a Nobel Prize. If you do some in-depth reading about the history of vitamins, you'll come across many of these arguments. As we continue with our story of vitamins, we will not be able to mention all the scientists who devoted their lives to this area of research.

The Nobel Prize was established by Alfred Nobel, inventor of dynamite, who gave his fortune to this cause.

In 1912, a Polish researcher named Casimir Funk was also working on the problem of beriberi. He isolated a chemical he thought was the anti-scurvy cure, and named it "***vitamine***," meaning "vital amine." (Remember, "amine" is NH_2.) What he had actually discovered was later found to be niacin (vitamin **B3**), unrelated to beriberi, but his discovery was still important.

In 1913, a researcher at the University of Wisconsin, USA, announced his discovery that these essential "vitamines" were of two types: "fat-soluble type A," and "water-soluble type B." This was the beginning of the long list of "B" vitamins. (B1 through B12)

Nicotaminide, a form of B3.

By 1920, other "vitamines" had been found that did not have an "amine" group (NH_2), but by now the name "vitamine" had become so popular there was no changing it. A British scientist suggested that everyone should start spelling it "***vitamin***," without the final "e," as a way to acknowledge the fact that many vitamins were not "amines." It's been "vitamin" ever since.

Riboflavin was discovered in 1922 in Germany and was eventually named vitamin **B2**. The same group of scientists went on to discover pyridoxine (**B6**) in 1934. During this time, an American pharmaceutical research company, Merck, decided that vitamins might turn out to be something that could be manufactured and sold to the public. Their first product was thiamine tablets for the prevention of beriberi. Then they jumped into the vitamin discovery race and their researchers announced the discovery of pantothenic acid (**B5**) in 1940.

B6 PYRIDOXINE

B2 RIBOFLAVIN

One of the last B vitamins to be discovered is cobalamin, **B12**. All during the 1800s and early 1900s, physicians were mystified about a rare diseased they called "pernicious *(per-NISH-us)* anemia." (Pernicious means "causing damage in a gradual way that is not easily seen." It is not a science word, and you might come across it in another context.) Patients with this illness would be tired and weak and short of breath. Blood tests revealed that too few red blood cells were being made, but the doctors did not know why. In 1926, scientists at Harvard University, USA, discovered that they could cure this disease by having patients eat half a pound of liver every day. Finding the critical substance in the liver was going to be difficult, however, because this type of anemia was never seen in animals. Without lab rats (or chickens or guinea pigs) to work with, researchers were limited in what they could do. Then, one day, an American microbiologist named Mary Shorb discovered a certain type of bacteria that seemed to react to liver chemicals the same way that human bodies did. Merck hired Shorb to assist their team of researchers who would attempt to isolate the curative agent in liver.

Fried liver and onions with potatoes

The liver chemicals that were most effective on the bacteria were pinkish in color, suggesting to the scientists that the mystery substance they were looking for was either pink or red. Sure enough, when they finally isolated this chemical, it came in the form of red crystals. They named it cobalamin, with the "cobal" part of the world coming from "cobalt," element number 27 on the Periodic Table, which was part of this vitamin's molecular structure. Finding cobalt in a vitamin was very surprising. Previously, cobalt was only known as a metal that was used to make magnets or to add blue (not red) color to glass. This illustration shows some other foods, besides liver, that contain significant amounts of B12.

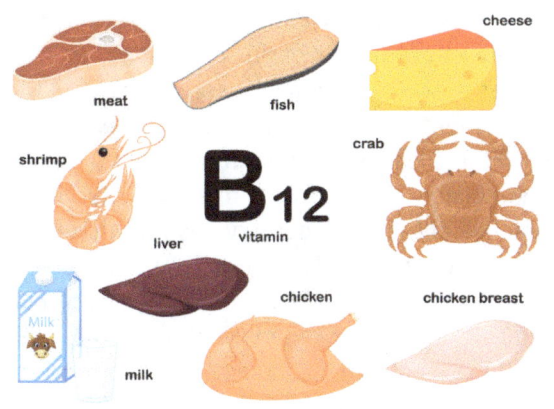

Well, we were on the topic of B vitamins, so it was best finish that topic in this chapter even if plants are not a good source of vitamin B12. In fact, before vitamin B12 pills were available, being a vegetarian (vegan) put a person at risk for pernicious anemia. However, very few people were vegan back then. For those who were, nutritional yeast was used as the primary source of B12. Now vegans can simply walk into any grocery store and buy a bottle of B12 tablets.

Before we leave the topic of vitamins, let's do vitamin D, too, since its discovery was linked to vitamin A.

Vitamin D was discovered by the same research team that discovered vitamin A. (You'll remember from chapter 3 that the vitamin A molecule is actually one half of a beta-carotene molecule.) They had discovered vitamin A in cod (fish) liver oil, and had also found this oil to be an effective treatment for a disease called **rickets**. People with rickets often looked "bowed legged" (bent legs) because the disease affects the ability of bones to maintain their hard mineral content. The researchers attempted to find out exactly what it was in the cod liver oil that cured rickets. Their first guess was that it was vitamin A itself, so to test their theory they took the vitamin A out of the oil and tried it on dogs who had rickets. The dogs still got better, even without the vitamin A in the oil, so they knew it must be a new vitamin not yet discovered. Even though they did not know exactly what it was, they still gave it a name. The obvious choice was "vitamin D" because we already had A, B and C.

COD LIVERS (Yum, yum.)

71

A few years later, other researchers discovered that vitamin D can be produced naturally by skin cells when they are exposed to ultra-violet light. In places of the world where people are out in the sun a lot, no one has rickets. Food chemists found that when they exposed food products such as milk, to UV light, the level of vitamin D went up. The use of UV light was an inexpensive and efficient way to fortify milk with vitamin D. Once this became common practice in the food industry, cases of rickets became extremely rare.

Vitamin E, discovered in 1922, is not a single molecule but a group of similar molecules. Plants make these molecules to catch dangerous "free radicals" (bits of broken molecules) that are created as UV light from the sun hits their leaves. Vitamin E molecules sit in the membranes we looked at on page 65 and protect those fancy little machines by catching and trapping free radicals.

By the mid-1900s, it had become obvious that these "vitamins" were absolutely essential to good health. Manufacturers of animal feed began putting vitamins into their products. Governments started telling companies that processed grains for human consumption that if their processes took out vitamins, they had to find a way to put them back in. When you see the word "fortified" on a product (often flour or cereals) this means that vitamins have been added in order to restore the original vitamin content of the grains.

VITAMIN K

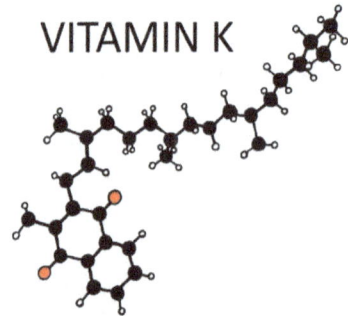

Before we compare vitamin contents of your salad greens, we need to mention a vitamin that most of these greens are high in: *vitamin K*. This vitamin comes in two forms. The one found in your salad leaves is called **phylloquinone**. You don't have to pronounce it or remember it. It's just interesting to see the proper names of these vitamins. See that long string of carbon atoms at the end? By now, you should easily recognize that this is what a fatty acid looks like, and you should be able to predict that this vitamin is in the "fat-soluble" category. That carbon tail will want to hang out with fatty acids which are also made of long strings of carbon atoms.

Your body uses this vitamin for several things, with the most well-known function being its participation in the blood clotting process. Vitamin K is necessary for making one of the proteins your body uses to make cuts and scrapes stop bleeding and form scabs.

Like vitamin K, many vitamins act as "cofactors" or "coenzymes" in body processes. Enzymes work on what is called the "lock and key" principle. The shape of the enzyme must match the shape of the substrates. There are a few enzymes whose natural shape isn't quite right, and they need another molecule (sometimes a vitamin) to fit into a gap, making the shape perfect. Vitamins are often the necessary "missing piece" needed to complete the shape and allow the enzyme to do its job.

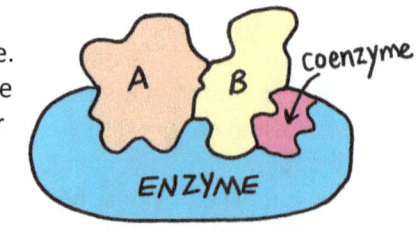

Now... let's stage a contest between your salad greens. We did the research for you (you're welcome). Use this chart to compare the levels of vitamins and minerals in four types of greens commonly found in salads. Of course, there would be phytonutrients in the leaves as well, but it is next to impossible to find enough detailed information on this to include it in the chart. In general, the darker the leaf, the more phytonutrients it has. You might already know what "fiber" means, but we'll discuss it in molecular detail on the next page. We also included some minerals that you may be familiar with. (Minerals are elements that you can find on the Periodic Table.)

You might have a different name for these greens. For example, "iceberg" is the most common name in the U.S., but in other places it is called "crisphead."

Most of the numbers in the chart are listed as percentages of the **Recommended Daily Allowance (RDA)**. You aim get at least 100% of every type of nutrient. So if one food gives you 5% of a vitamin, that means you still have to get the remaining 95% of that vitamin from the other food you eat that day.

NOTE: These numbers are for serving size of 100 grams (3.5 oz).

	ICEBERG	ENDIVE	RED LEAF LETTUCE	SPINACH
sugars	1.97 g	.45 g	.48 g	.4 g
fiber	.2 g	3.1 g	.9 g	2.2 g
vit. A	3%	72%	47%	59%
thiamine (B1)	2%	7%	6%	7%
riboflavin (B2)	2%	6%	6%	16%
pyridoxine (B6)	2%	1.5%	8%	15%
folate (B9)	7%	36%	9%	49%
vit. C	5%	8%	4%	34%
vit. K	20%	220%	134%	460%
calcium	2%	5%	3%	10%
iron	1%	6%	9%	21%
magnesium	2%	4%	3%	22%
phosphorus	3%	4%	4%	7%
sodium	1%	1.5%	2%	5%
zinc	1%	8%	2%	6%

And now, fiber...

Technically, *fiber* isn't really a nutrient. In fact, for the most part, it goes through your digestive system intact and undigested. Some people called it "roughage." Before the late 1900s, fiber was under-appreciated as a necessary part of a healthy diet. If you can't digest it, then what good is it, right? However, scientific studies began collecting data on people whose diets contained a lot of fiber and on those who had very little. The results of the studies showed that people who ate lots of fiber had healthier digestive systems as well as having fewer problems with other internal organs such as heart and liver. It became obvious that eating plants that you can't digest is actually good for you.

At first, scientists speculated that the fiber was causing the food to move along more quickly through the intestines, preventing the build-up of toxins along the way. They also guessed that perhaps the bulk of the fiber was somehow mechanically "cleaning" the inner walls of the intestines. Both of these theories are likely to be true, but there was a key element they had overlooked.

At the very end of the 1900s, a whole new branch of science began: the study of the **microbiome**. "Micro" means "small," and "bio" means "life." Your microbiome consists of all the microscopic organisms that live inside of you. Most of them are beneficial, not harmful, and you really can't live without them. They keep "bad" bacteria out, they make helpful chemicals such as vitamins (B7 and B9) and fatty acids, and they even play a role in brain health. The number of bacterial cells in your body is ten times the number of your own cells! Much of your microbiome is found in your intestines, particularly the large intestine. Like all living things, these bacteria have to eat, and many of them love to eat the plant fibers that you can't digest. As a general rule, the species of bacteria that are the least helpful are the ones who like to eat sugar and "junk food," and the species that are most helpful are the ones who like to eat healthy food such as vegetables and salad greens.

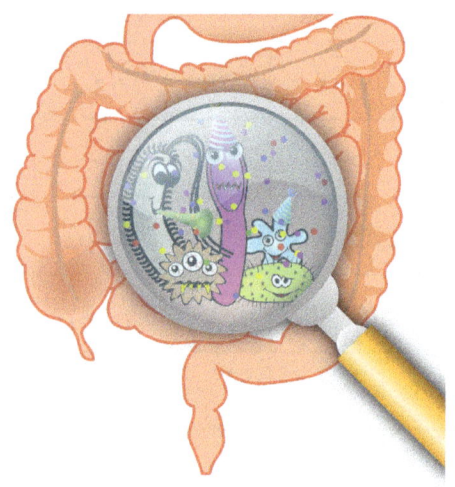

Fiber is made of long molecules called **cellulose**. At first glance, cellulose molecules might look identical to starch. However, if you look carefully at the orientation of the glucose units, you will see the difference. In starch, all the glucose molecules are sitting the same way, with their "flags" (CH_2OH) at the top. In cellulose, every other glucose molecule has its "flag" pointing down: up, down, up, down, up, down, etc. This is a small difference, but it has a large effect. Our bodies can only tear apart glucose molecules that are all facing the same way.

STARCH

CELLULOSE

Enzymes, as we have mentioned previously, can only do one job. The enzyme that can disconnect the glucose molecules found in starch has a shape that fits exactly into the space between the glucose units. In this illustration, we've drawn this enzyme as a blue shape that acts like a key. (The real enzyme doesn't look anything like this, but the blue shape will help us to understand the function of the enzyme.) The enzyme that is needed to disconnect the glucose units in cellulose is shown in orange. You can see that the keys have different shapes.

Our bodies don't make those cellulose keys; we only make starch keys. Bacteria, however, do make the cellulose keys, so they are able to tear apart cellulose and use the glucose molecules.

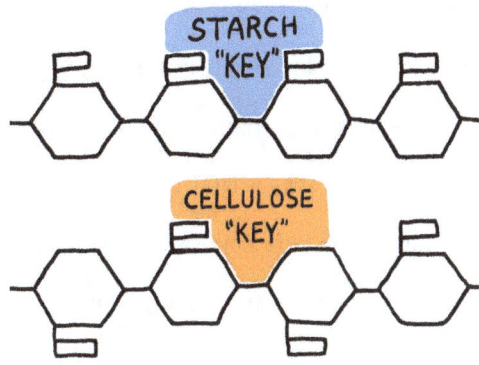

The correct name for the cellulose enzyme key is *cellulase*. It's not just humans who can't manufacture cellulase—not a single mammal can. Grazing animals like cows and sheep can live on nothing but grass because their digestive tracts house massive numbers of bacteria. Cows have a collection bag where all the grass goes, and in this bag are not only bacteria but also some one-celled protozoans. These microorganisms feast on all the grass the cow has swallowed. The semi-digested grass, and many dead bacteria, then go into a regular stomach where digestive enzymes finish the job and prepare the nutrients to be absorbed in the intestines.

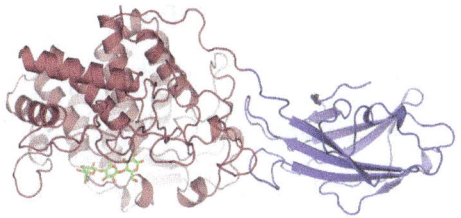

CELLULASE

By B.T.Riley - Own work, CC BY-SA 4.0, https://commons.wikimedia.org/w/index.php?curid=46569786 rid=46569786

Let's end by summarizing our salad. What is salad made of? Over 90 percent of it is water. If you dehydrated your salad it would shrivel down to almost nothing. The next most abundant ingredient is cellulose, which makes the thick cell walls around all the cells. In third place we'd have a tie for sugar (in the form of either starch or free floating glucose molecules) and protein (which would include all those fancy little machines that do photosynthesis). The remaining small percentage would be nucleic acids (DNA and RNA), fats, vitamins, minerals, pigment molecules (green chlorophylls in lettuce, orange beta-carotenes in carrots, and red anthocyanins in tomato) and other phytonutrients.

(Don't forget to check out the videos that go with this chapter at YouTube.com/TheBasemenetWorkshop.)

Comprehension self-check

1) Water is polar because the _____ are shared unevenly.

2) An emulsion is made of a polar substance and a non-polar substance held together by an _____.

3) Which of these is NOT used as an emulsifier in dressings?   a) egg yolks   b) salt   c) mayonnaise   d) mustard

4) How many fatty acid chains are on a diglyceride molecule?____   A monoglyceride? _____   A triglyceride? ____

5) To a scientist, a fruit is anything with _____. Which of these would a scientist NOT classify as a fruit?
   a) cucumber   b) green pepper   c) sweet potato   d) tomato

6) The tiny holes in leaves are called _____, which are made of two _____ cells.

7) Which layer in a leaf contains the greatest number of chloroplasts?

8) TRUE or FALSE? Plant DNA is fundamentally different from human DNA.

9) The membrane around a cell, or around a thylakoid disc, is made of "2-legged" molecules called:

10) Which pigment molecule has an "X" in the middle of a square shape?

11) Which pigment molecule gets chopped in half to make vitamin A for your eyes?

12) The square part of chlorophyll is very similar to this molecule in your blood that carries oxygen:

13) What is the atom at the very center of (the square part of) the chlorophyll molecule?

14) Lycopene and lutein belong to what category of phytonutrients?

15) What flavonoid molecule is responsible for red, blue and purple coloration in plants?

16) This bright yellow polyphenol phytonutrient is used in Indian recipes:

17) Substances that can capture dangerous "broken" molecules and single oxygen atoms are called:

18) Tea leaves contain this tangy phytochemical:

19) From a plant's point of view, what are phytochemicals for?

20) What disease is caused by a lack of vitamin C? _____   Vitamin B1? _____

21) What does the "-amin" ending mean, in the word vitamin?
   a) It is an acid.   b) It is a base.   c) It has a COOH group.   d) It has an NH_2 group.

22) Which one of these is NOT a B vitamin?   a) resveratrol   b) thiamine   c) pyridoxine   d) riboflavin   e) cobalamin

23) Which one of these is NOT a good source of vitamin B12?   a) chicken   b) liver   c) spinach   d) fish   e) cheese

24) Name the two vitamins that were discovered in cod liver oil:

25) Which vitamin can be made by your skin cells when exposed to ultraviolet light?

26) Which vitamin is needed for the blood clotting process?

27) What do you call a small molecule that is needed by an enzyme to complete its shape?

28) Fiber is made of c_____.   The enzyme needed to tear it apart is called c_____.

29) Can any animals digest fiber?

30) Why do you need to eat fiber?

These question use the chart on page 73.

31) Which type of salad green has the greatest amount of vitamin K?

32) Which type of salad green has the greatest amount of natural sugars?

33) Which type of salad green has the highest level of iron?

34) Which type of salad green as the most fiber?

35) Like humans, guinea pigs cannot make vitamin C and must get all of it from their diet. Which type of salad green would be the best to feed to your guinea pig?

36) Which type of salad green has the most magnesium?

37) People who take medication to prevent blood clots are often told not to consume a lot of vitamin K. Which salad green would you recommend that they eat?

38) Which type of salad green has the least amount of B vitamins?

39) Many people like to be careful not to consume too much salt, which includes sodium. Would these people have to limit their intake of salad greens?

40) Which salad green contains the greatest amount of vitamin A?

ACTIVITY 4.1   BONUS INFORMATION: Anthocyanin— the amazing color-changing molecule!

Anthocyanins are a family of pigment molecules that share the basic structure shown here. They are in the polyphenol group and are manufactured by plant cells using the amino acids phenylalanine and tyrosine. After the molecules are made, they are put into a storage vacuole. Cells in any part of a plant can make anthocyanins: roots, stems, leaves, and fruit.

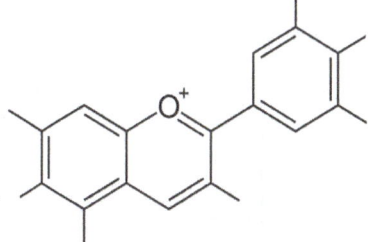

When anthocyanin molecules are in an acidic environment, the acid causes the molecule to reflect only red light. As the pH rises and the environment becomes less acidic, the molecule will change shape just enough that it no longer reflects red light but reflects blue instead. If the environment is neutral (7 on the pH scale) the anthocyanin will reflect dark purple light. We see these dark purples in eggplants and black beans. If the pH starts to become alkaline (higher than 7 on the pH scale) the molecule will stop reflecting purple light and begin to reflect green. In a very alkaline environment, the molecule reflects yellow light. The picture of the test tubes shows these color changes.

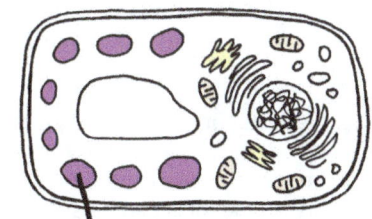

vacuoles full of anthocyanins

This ability of anthocyanins to change color explains why many berries, including raspberries and blueberries, are red when they are unripe, but blue or purple when ripe. Have you ever tasted an unripe berry? Wow, are they sour! When something tastes sour, it is acidic. As the berry ripens, the chemistry changes and the berry becomes sweeter and much less tart. The ripe berry tastes less sour because it is less acidic. This change in pH as the berry ripens affects both taste and color.

There's another chemical that can make plants red, called **betalain**. Plants that produce betalain don't make anthocyanins. Betalain is found in beets, Swiss chard, and many cactus flowers. Betalain doesn't change color; it stays red regardless of the pH around it.

By Indikator-Blaukraut.JPG: Supermartlderivative work: Haltopub (talk) - Indikator-Blaukraut.JPG, CC BY-SA 3.0, https://commons.wikimedia.org/w/index.php?curid=7741760

ACTIVITY 4.2
SALAD SCIENCE
CROSSWORD

ACROSS:
1) This group of polyphenols gives tea its tangy taste.
3) The scientist that gave us the word "vitamin."
5) The scientific name for vitamin B2
6) This yellow spice contains the polyphenol curcumin.
11) This word means "water fearing/hating."
13) This condition is caused by lack of vitamin D.
14) The _____ layer in a leaf has many air spaces.
15) The process of turning sunlight, water and carbon dioxide into glucose (plus oxygen and water).
17) A small molecule that completes the shape of an enzyme.
18) Anything with a seed is technically a _____.
23) A substance that can hold both polar and non-polar molecules.
24) The scientific name for vitamin B6
26) This polyphenol is found dark red or blue fruits and has anti-inflammatory properties.
27) This very long molecule is a primary building material plants use to make their cell walls.
28) Any molecule or substance that can catch and trap dangerous molecules, such as lone oxygen atoms.
29) The empty "bubble" in the middle of a plant cell.

DOWN:
2) A string of glucose molecules that can be broken apart by the enzyme amylase.
3) The general word describing any substance that goes through our intestines undigested.
4) The scientific name for vitamin B1
7) The term used to describe all the bacteria that live in and on you.
8) The organelle in a plant cell where you find chlorophyll molecules
9) Phytonutrients that have in their molecular structure at least one hexagonal ring of carbon atoms with an OH attached.
10) The name of the molecule that forms membranes
12) Vitamin C is also called _____ acid.
14) The microscopic holes in a leaf
16) Phytochemicals that reflect orange light
19) The waxy surface layer of a leaf
20) The molecule in our blood cells that carries oxygen molecule (similar structure to chlorophyll)
21) The enzyme that can break apart cellulose
22) The atom at the very center of the chlorophyll molecule, where photons of light are captured
25) DNA and RNA are _____ acids.

ACTIVITY 4.3 Fourth installment of "Chew It Over," a group game to be played during a meal

Here is another round of questions for you to use at a mealtime that you share with family or friends. These questions relate to the topics we learned about in this chapter. Again, you can use these questions in a variety of ways. You can be the quiz master and determine who gets which questions, or you can cut the questions out of the book and put them into a bag or bowl and let people choose a question randomly. The answers on are the back of this page.

CHAPTER 4: SALAD

1) Name a plant that you like to eat if it is cooked, but you don't like if it is raw. Then name one you like raw but not cooked.

CHAPTER 4: SALAD

2) Guess what percentage of the population in America admits that they almost never eat vegetables.

CHAPTER 4: SALAD

3) Guess which root vegetable is reportedly the most hated (according to a U.S. survey)?

a) beet   b) turnip
c) rutabaga   d) radish

CHAPTER 4: SALAD

4) All of these vegetables have purple varieties except one. Which one is never purple?

a) carrot   b) cabbage   c) potato
d) cucumber   d) cauliflower   3) pepper

CHAPTER 4: SALAD

5) Guess which of these scores higher on the favorite vegetables list (according to a U.S. survey): Green beans or green peas?

CHAPTER 4: SALAD

6) Guess which U.S. state holds the record for growing the largest vegetables.

a) Alaska   b) Hawaii
c) Texas   d) Florida

CHAPTER 4: SALAD

7) Four of these vegetables belong to the same family of plants, commonly called the "nightshades." Can you guess which one is not related to the others?

a) potato   b) pepper   c) eggplant
d) avocado   e) tomato

CHAPTER 4: SALAD

8) STRANGE FACT:
The Vegetable Orchestra, based in Austria, makes all their instruments from fresh vegetables. They assemble the edible instruments before each performance, then serve vegetable soup to the audience afterwards.

CHAPTER 4: SALAD

9) How many salad greens or vegetables can you name that are <u>always</u> eaten raw and never cooked?

CHAPTER 4: SALAD

10) INTERESTING FACTS ABOUT CUCUMBERS:
1) The skin of a cucumber can erase pen marks.
2) A slice of cucumber can be used as a breath freshener.
3) Cucumber juice has been used for waterproofing.

CHAPTER 4: SALAD

11) Go around the table and have each person try to name five foods they have eaten today or yesterday that contain some kind of phytonutrient. (You don't have to name the phytonutrients.)

CHAPTER 4: SALAD

12) If you had to take a job working with lab animals, which one would you choose?

a) mice   b) rats
c) guinea pigs   d) chickens

2) 25%
3) b) turnip
4) d) cucumber
5) Green beans
6) Alaska, because in the summer they get up to 20 hours of sunlight each day.
7) d) avocado (Yes, the potato and tomato are related!)
9) cucumber, lettuce (though lettuce is cooked in some parts of the world), radish, perhaphs you know another

5: Main course

This type of main course, with both fish and red meat, is sometimes called "surf and turf." Fish live in the ocean, which has waves you can surf. Turf is a word that means a surface of grass and dirt, like a field used for grazing cows, and cows are the main source of red meat. Actually, a better name for your plate might be "Surf & Turf Plus," because we've added a third feature we'd like you to try: a newly invented plant-based meat substitute.

But first, let us direct your attention to that little bottle. We've provided you with some hot sauce in case you want to spice up your meat or potatoes. If you don't like having your mouth feel like it is on fire, you are welcome to say, "No, thank you." Regardless of whether you use the hot sauce or not, we'll take a look at the molecular structure of the chemical that makes hot sauce hot, and learn why it works.

First, we need to look at a microscopic structure found in the outer membrane of your nerve cells. You'll remember that we saw lots of little gadgets (made of protein) in the membranes of the green thylakoid discs inside plant cells. Some acted like pumps or shuttle buses and the biggest one was a motor that recharged ATP batteries. All cells have lots of little gizmos and gadgets embedded in their outer membranes. Some of these gadgets act as portals that control the entry or exit of certain molecules.

Here we see in our magnifier a portal with a very long name that we won't bother you with. These portals are found in your nerve cells and their job is to alert you to danger. When triggered, the portal will allow a sudden

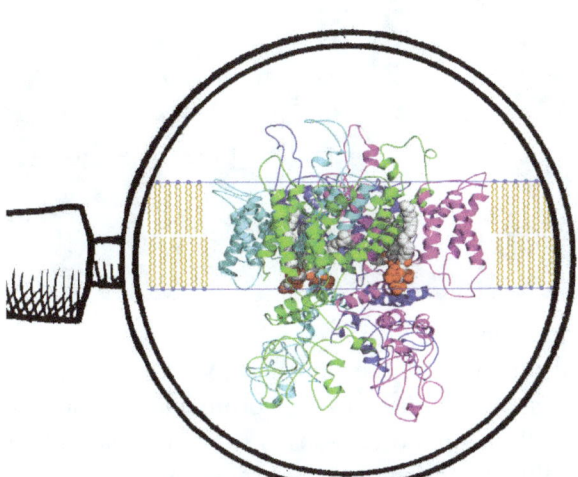

inflow of calcium atoms into the cell, which will cause the cell to generate an electrical signal that will eventually be sent to the brain. The brain will interpret this particular signal as heat. If the signal is very strong, your attention will be drawn to where the signal is coming from in case you need to do something like get your hand away from the source of the heat.

This portal can be triggered by a number of things, including: 1) temperatures above 109°F (43°C), 2) the presence of many protons (usually from an acidic substance), 3) a chemical that is found in mustard and in wasabi, and 4) a chemical called **capsaicin** (cap-SAY-sin) which is found in hot peppers. No matter what triggers the portal, the same signal will be sent to the brain: "HOT!"

Here is the capsaicin molecule. Something about its shape is just right to trigger that portal. There's isn't anything "hot" about the molecule. It's just a molecule. But it triggers that portal and a signal is sent to the brain, and the brain interprets the signal as a hot sensation. Most mammals have this same portal protein in their nerves, so hot sauce is sometimes recommended as a non-toxic way to deter wild animals from pesky behavior. For example, some people mix hot pepper seeds into the seeds in their bird feeders because squirrels are sensitive to capsaicin while birds are not. (Although the author of this book tried to use triple-hot pepper sauce to keep her border collie out of the trash and it didn't work.)

From the plant's point of view, the capsaicin is a defensive mechanism, to deter bugs and small animals from eating its fruit and seeds. The plant needs its seeds so it can make baby plants. And speaking of seeds, let's examine your green beans next. You can use your knife to cut one open.

Some of your beans may have very small seeds, but a few will have seeds large enough for us to work with. We'll cut them open in a minute.

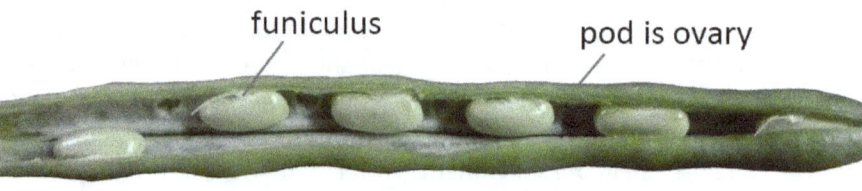

Bean pods are technically the plant's **ovaries**. In plants, the ovary is the structure that contains the seeds. The part labeled "funiculus" is where the seed is attached to the ovary. The pod started out as a tiny ovary at the base of a bean flower. After the flower was pollinated by an insect, the seeds started to grow inside the ovary. The ovary then grew into the long green pod we recognize as a bean. We like to pick bean pods while they are still growing, well before they mature. Once they mature and start to ripen, the pods become tough and lose their sweet taste, and the beans become hard and chewy.

As you eat your beans, you can be thankful for the work of a man named Calvin Keeney who lived in New York in the late 1800s. Until Calvin started breeding bean plants, bean pods had a very prominent "string" running along the seam on one side. (This is why many people still call them "string beans.") The string wasn't very pleasant to chew and could get stuck between your teeth. It's amazing that people ate them anyway.

Calvin found bean plants that had slightly less stringy pods. He would breed less stringy plants with other less stringy plants and they would produce plants that were even less stringy. He kept breeding until he had produced beans with a "stringless" pod. The Burpee seed company began marketing the stringless beans in 1894 and they remained the best selling bean in the world until the Blue Lake variety was developed in the 1950s.

If we carefully slice open one of the bean seeds, we can see its anatomy. There is actually a tiny baby plant inside! This plant embryo already has two tiny leaves called *plumules*. Below the plumules is the part that will turn into stem and roots. The bulk of the seed is called the *cotyledon* (cot-ill-L EE-don), also known as the *seed leaves*. Bean

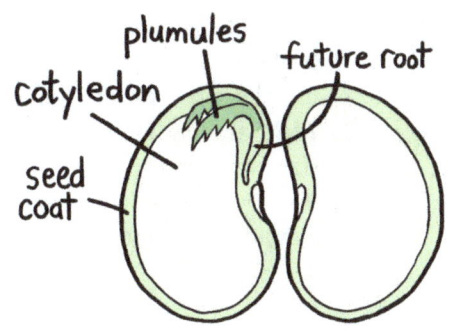
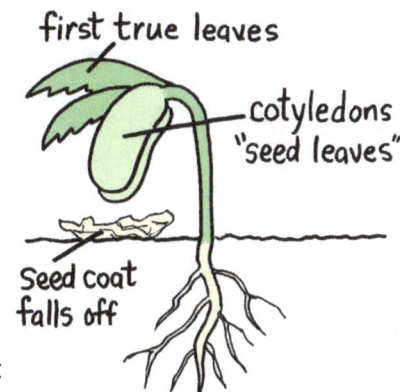

seeds have two cotyledons, which is why they are so easy to open; there is sort of a natural split in the middle. The seed has a covering around it, called the *seed coat.* The seed coat contains a lot of cellulose to make it tough, and extra phytochemicals to discourage bacteria and bugs from entering.

We are going to focus on the cotyledons, since they form most of the seed and therefore give it its nutritional properties. (The seed coats adds fiber and minerals such as zinc, calcium and magnesium.) The function of the cotyledon is to provide energy for the baby plant before it develops enough leaves to begin using photosynthesis to make its own food. The energy is mostly in the form of starch molecules.

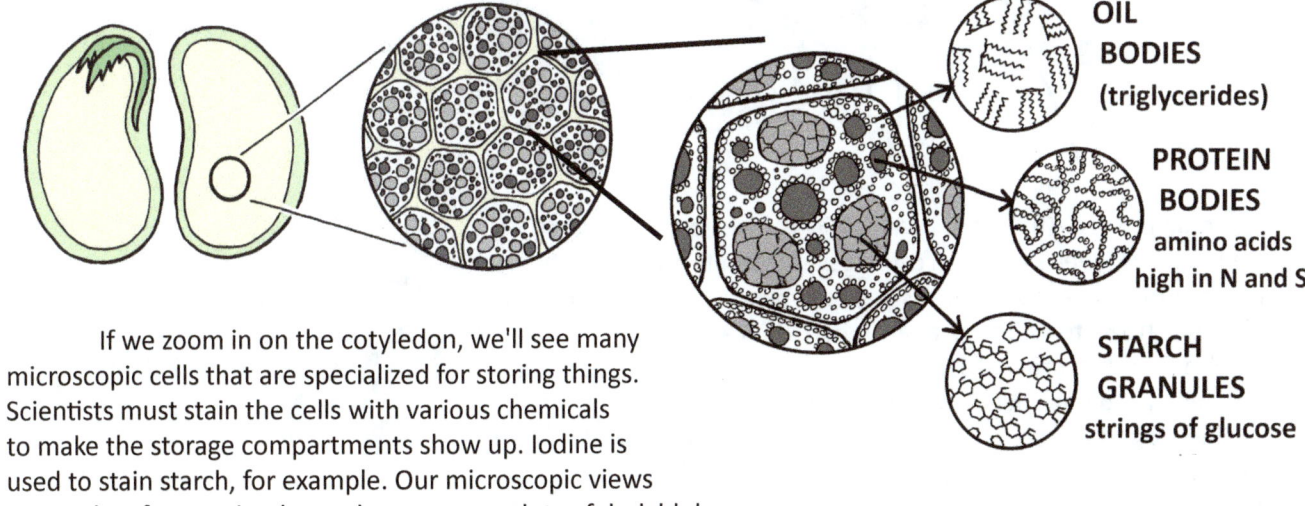

If we zoom in on the cotyledon, we'll see many microscopic cells that are specialized for storing things. Scientists must stain the cells with various chemicals to make the storage compartments show up. Iodine is used to stain starch, for example. Our microscopic views were taken from stained samples, so we see lots of dark blobs.

There are three types of storage vacuoles: oil bodies, protein bodies, and amyloplasts.

Oil bodies are tiny spheres made of membrane that are filled with *triglyceride* molecules. We've seen triglycerides several times now, so you will remember that they are made of a "hanger" molecule (glycerol) with three fatty acid "tails" made of strings of carbon atoms with hydrogens attached. We often represent these carbon chains as zigzaggy lines. The length of the carbon chains will vary, usually be between 14-22 carbons long. The oil bodies are very small and tend to cling to the outside of the protein bodies or around the edges of the cell.

Protein bodies, which are shown as dark circles in our picture, are also surrounded by membrane and are filled with strings of *amino acids*, especially aminos that have **sulfur** atoms or extra **nitrogen** atoms. The baby plant will need a source of nitrogen and sulfur in order to grow.

Amyloplasts store starch granules that are made of strings of glucose molecules. The length and shape of the strings differ from seed to seed. Some starch granules have a smooth appearance and other seeds make their granules with geometric shapes, looking a bit like turtle shells.

From the plant's point of view, the oil, protein, and starch are meant to feed the baby plant. Baby plants need a balanced diet with plenty of nutrients, just like you do.

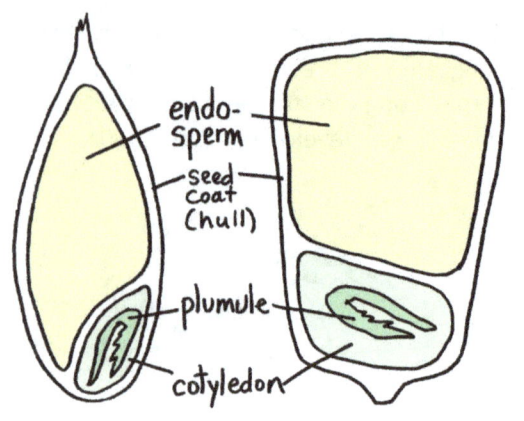

Rice and corn seeds are a bit different from bean seeds. They have only one cotyledon, not two. The bulk of a rice or corn seed is called **endosperm**. Like bean cotyledons, rice and corn endosperms store oil, protein and starch. Rice and corn tend to have more starch than beans do. The starch molecules in rice can be straight or branched, as we learned in chapter 3. In wheat seeds, the endosperm is where you find the proteins glutenin and gliadin. They would be inside protein bodies.

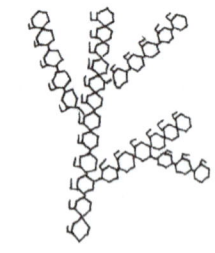

STARCH (strings of glucose)

Actually... yes. The entire bean family is toxic to some degree. Eating a few raw kidney beans can cause severe vomiting and diarrhea. Eating a handful of raw kidney beans will put you in the hospital. Green beans are much less toxic. Eating a raw green bean as you walk through your garden isn't dangerous. A large quantity of raw green beans could give you quite a stomach ache, though.

The toxic chemical in beans is called **phytohemagglutinin** *(fie-toe-hee-mah-glue-tin-en)* and it belongs to a large family of molecules called **lectins**. Lectins are proteins so they are made of amino acids. Here on the right you can see how amazingly beautiful this protein is. Of course, the colors have been added, but even without the color it's still nice-looking.

If we look closely at the name of this toxin, we see "hem" in the middle, which might remind us of the word "heme" from the last chapter. Heme is the molecule in our red blood cells that carries oxygen. At the end of the word we have "agglutin" which comes from the world "agglutinate," meaning "to clump together." Wait... this sounds bad. Heme molecules clumping together? That would be nasty. And what does this have to do with bean seeds?

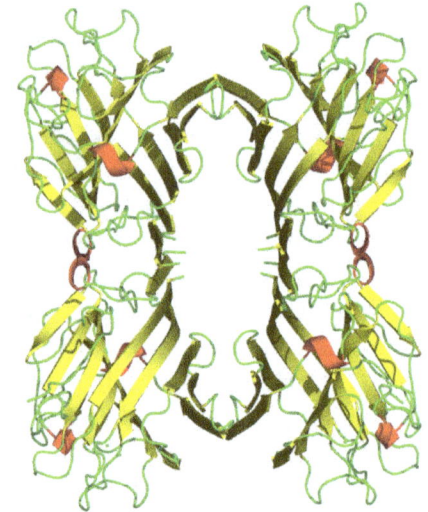

PHYTOHEMAGGLUTININ
Beautiful but dangerous.

Lectins are not bad molecules. They are found throughout all kingdoms of life, and even in some viruses. They are proteins that are designed to bind to a specific type of carbohydrate molecule (made of sugars) or glycoprotein molecule (combination of sugar and protein). In human bodies, various types of lectins perform essential tasks such as promoting bone growth or regulating the amount of certain proteins in our blood. They play a role in our immune systems, helping our white cells to recognize foreign invaders.

What lectins do in plants remains somewhat of a mystery. They occur mostly in seeds, so we can guess that they play a role in germination and sprouting of seeds. Botanists guess that lectins might assist in communication between parts of a plant, or between plants, but we don't yet know how it happens.

The proteins in phytohemagglutinin just happen to be the right shape to stick to little sugar strings that are on the surface of your blood cells; they act like a ball of sticky tape, holding on to multiple blood cells. (However, ingested raw beans are likely to be vomited up long before they are digested enough to get into your blood. This connection to blood is mostly a laboratory observation, not something that happens in your body.) Additionally, phytohemagglutinin can cause cells to suddenly decide to divide in half in a reproductive process called mitosis. And if that were not enough, some lectins can interfere with your ability to absorb minerals such as iron, zinc and calcium. (For this reason, lectins have been called "anti-nutrients" and have been the target of fad diets that give you a long list of things you shouldn't eat.) All these bad traits can be put to good use, however, by research scientists. Hemagglutinin molecules are a valuable tool that researchers can use to study blood and to learn about certain cellular processes.

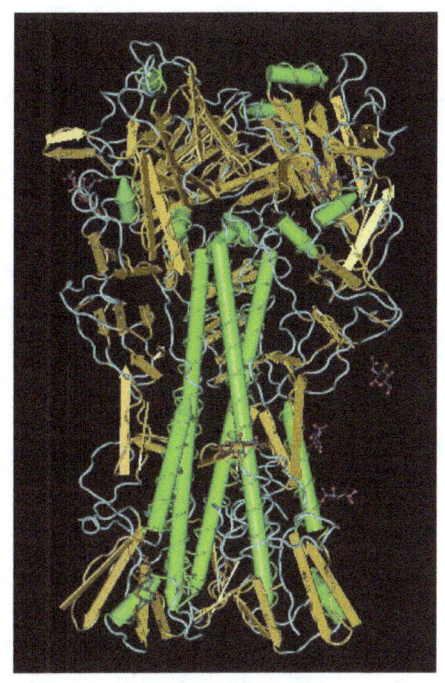

HEMAGGLUTININ

The good news (for those of us not involved in research) is that the toxicity of lectins can easily be neutralized by cooking the beans. All of the beans you buy in the store will have been washed and pre-cooked so there is no danger in eating them. Even bags of dry beans have been pre-cooked. The beans were soaked, rinsed, and then boiled at a high temperature for at least 30 minutes. There are many health benefits to eating cooked beans. They are high in fiber, protein, and minerals. Some people groups around the world eat beans almost every day, and stay healthy doing so.

Undoubtedly you are referring to a poem whose punch line is funny because it involves intestinal gas. The scientific word for the production of intestinal gas is "flatulence," and yes, beans can do that, but the molecule that causes this problem is not a protein or a phytochemical. The offending substance is called **raffinose**, and is found in other plants, also, not just beans. It is a small molecule made of only three simple sugars. Back in chapter one we learned about sucrose and saw that it is a disaccharide, made of glucose and fructose. Raffinose is a trisaccharide made of three simple sugars: glucose, fructose and galactose. If you stick a galactose onto a sucrose molecule, you get raffinose. You'd think that since our bodies can digest lots of sugars and starches, it wouldn't be a big deal to clip off that galactose.

85

However, as we've mentioned previously, enyzmes are very specific to a certain task and, as a general rule, can only do one thing. So you need the right enzyme to clip off that galactose and turn the molecule into sucrose (table sugar) that we easily digest. The name of this enzyme is **alpha-galactosidase**. *(gal-ack-TOSE-i-dase)*

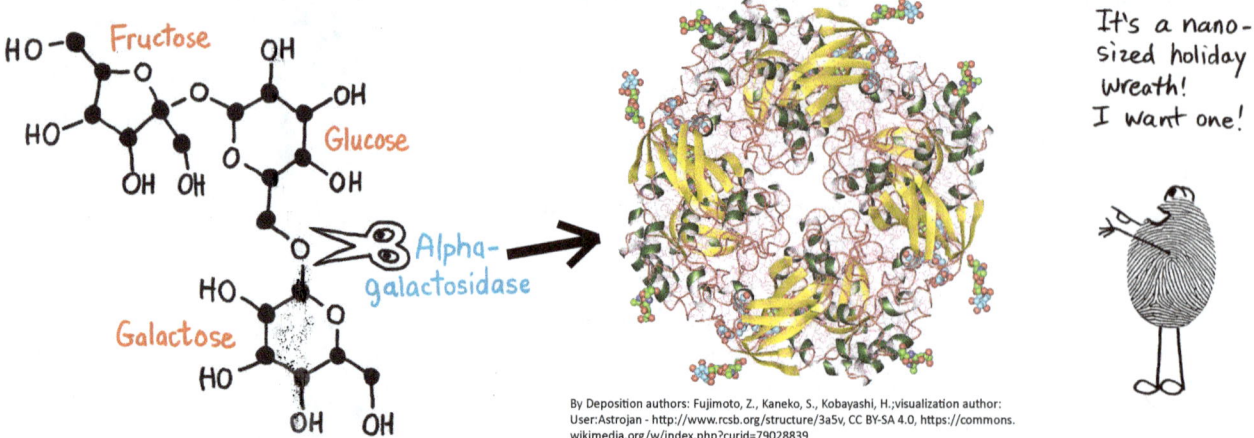

All those curls and folds are made of amino acids. The amino acids have to be in exactly the right order for this to happen. The instructions for the order in which to string the amino acids is found in DNA. Unfortunately, human DNA doesn't have the instructions for how to make this enzyme. The bacteria in your intestines, however, do have this information and therefore can make this enzyme. So when the raffinose gets to your large intestine, it's dinner time for your bacteria! As raffinose is torn apart, smaller molecules are created, including molecules of various gases. So it isn't your body creating the gas, it's your bacteria. If you would like to be able to digest the raffinase before it gets to the large intestine, you can take digestive aids, such as BeanO®, that contains alpha-galactosidase.

Humans have a version of alpha-galactosidase that can snip off a galactose sugar on the end of other short sugar strings, but not raffinose. There are lots of places on and in your cells where you'd find short sugar strings (oligosaccharides) and sometimes they need to have a galactose snipped off the end. One place you meet short sugar strings is on the outside of red blood cells. These oligosaccharides determine blood type (A, B, AB and O). Blood banks can use an alpha-galactosidase enzyme to turn type B blood into type O by snipping off a galactose. This is important because type O is the "universal donor" that you can give to anyone, regardless of their blood type. Ambulances carry only type O.

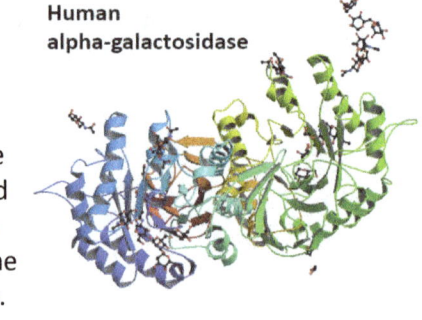

Yes. The bad news is that potatoes can also produce toxic chemicals. Fortunately, the potato plant produces a lot of chlorophyll alongside the toxins, so the poisonous areas are bright green and easy to see. The toxin is called **solanine** (sole-uh-neen). Potatoes, and other plants in the "nightshade" (*Solanum*) family, such as peppers and tomatoes, produce solanine as a natural pesticide. If you are a bug, a bite of green potato could be lethal. We are a lot bigger than bugs, so it takes more than one bite to harm us. However, cases of solanine poisoning do happen, so it is always a good idea to peel off any green parts on a potato, and check to see whether the white part tastes bitter.  If the white flesh is bitter, don't eat it. Symptoms of solanine poisoning include nausea, vomiting, headaches, dizziness, itching, heart palpitations, and in severe cases, paralysis. The solanine interferes with the functioning of the portals that let calcium atoms go in and out of nerve cells, so there's calcium flying around everywhere, making your nerve cells go haywire. Unlike lectins, solanine can't be destroyed through cooking, so boiling green potatoes doesn't solve the problem. You need to peel or cut off the green parts.

Now for the good news about potatoes. Potatoes have a lot of starch, which is a good source of calories, and they also contain many vitamins and minerals. If the skins are not green, they are a terrific source of fiber and phytonutrients. Potatoes will grow in northern climates, so they have been vital to the survival of people living at high latitudes, such as the British Isles, northern Europe, Scandinavia, and Russia.

Unlike potatoes, sweet potatoes do not grow well in northern climates. They are not in the nightshade family of plants, and therefore not closely related to regular potatoes. Sweet potatoes are members of the plant family to which the "morning glory" flower belongs. Like carrots, sweet potatoes are most often orange, but can also be white, yellow, or purple.

Morning glory flowers close in the afternoon.

Sweet potatoes are often mistakenly called **yams**. True yams are usually white inside, not orange, and they have a rough, brown exterior. Yams are mainly grown in Africa, whereas sweet potatoes are widely distributed around the world. Scientists have used DNA evidence from archaeological sites to determine that thousands of years ago ancient travelers took sweet potatoes from Central and South America across the Pacific Ocean to many island nations and finally to Japan and China. Much later, in the 1500s, Europeans who explored Central America took them back home to Europe. The word "potato" came from the name that the Taino people (on Caribbean islands) used for this food: "batata."

Since potatoes and sweet potatoes grow under the ground, it seems logical to assume that they are roots. Actually, neither of them is a root—they are both **tubers**. Tubers are modified plant parts (often stem or root) that the plant uses to store energy. Regular potatoes are **stem tubers** that develop on modified stems called **stolons**. Sweet potatoes are **root tubers**.

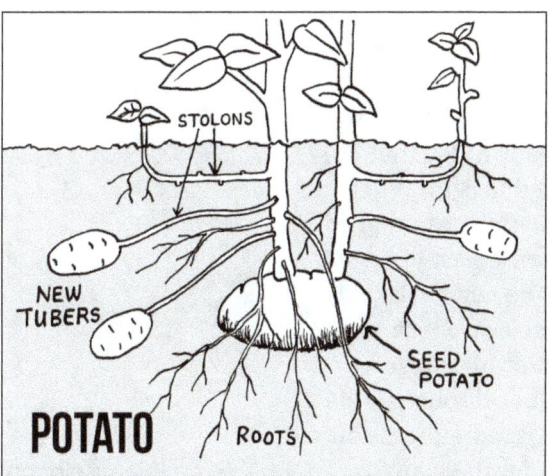
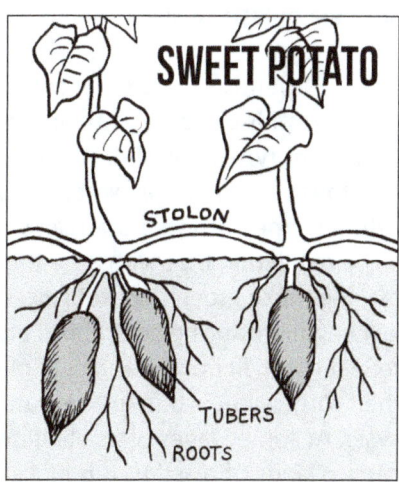

The cells of both kinds of tubers are filled with amyloplasts containing starches.

Now it is time to get out our Sooper Dooper magnifier and compare the cells of the potato and the sweet potato. Our magnifier will be used on its lowest setting, which is the same as a regular microscope in any high school or college lab. If you have a microscope at home, you can do this yourself. We will cut a very thin sliver from each kind of potato, then add a drop of iodine ("Lugol's solution") which will stain the starch molecules.

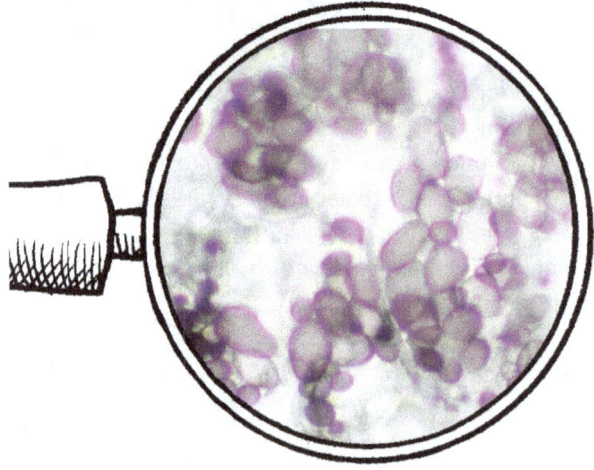 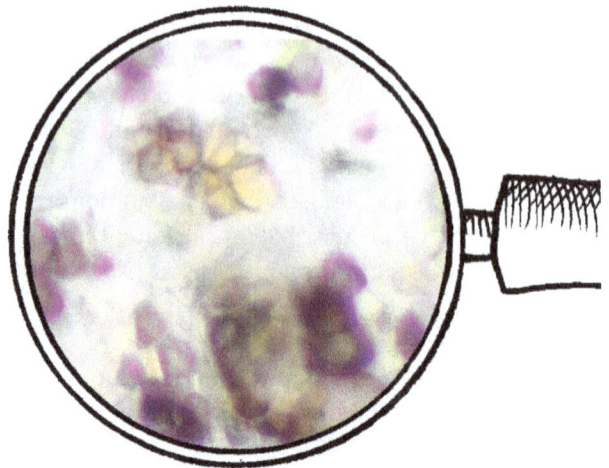

WHITE POTATO 100x   SWEET POTATO 100x

The purple ovals are the amyloplasts. The faint lines between the ovals are the cell walls. Each view shows about half a dozen cells. We can see that the sweet potato cells have orange ovals as well as purple ones. The orange ovals are called **chromoplasts**. They have not been stained; orange is their real color. "Chromo" means "color," so the chromoplasts are where you find a plant's pigment molecules.

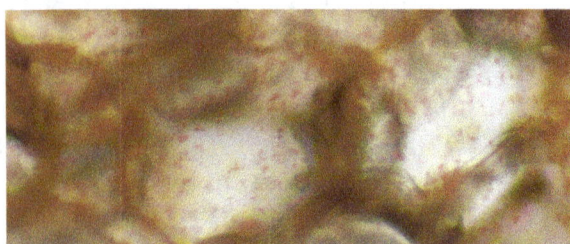 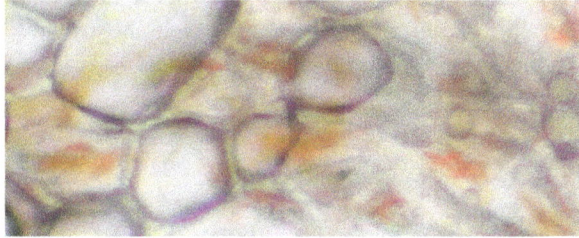

This is a thin slice of red pepper. The tiny red dots are the pepper's chromoplasts.

This is a thin slice of carrot. The orange chromoplasts are not as well defined and appear oblong.

Amyloplasts, chloroplasts and chromoplasts are **plastids**. Other plastids include **phenyloplasts**, which contain dark-colored polyphenols, and **elaioplasts**, which store fat molecules. The cells in the skins of citrus fruits have elaioplasts, which explains the term "citrus oil." Citrus oil is used in organic insecticides.

Scientists were surprised to discover that many plastids can change their function and become a different type of plastid. We've already noted that potatoes will turn green when exposed to light. When light hits one of these potato amyloplasts, it can trigger a chain reaction that will cause the amyloplast to begin making those green thylakoid discs. The new thylakoids will begin producing chlorophyll molecules, which reflect green light. The ability of plastids to change their form is important as the plant goes through its life cycle. At some stages of its life it needs more or less of certain kinds of plastids. It is efficient for the plant to "repurpose" its plastids.

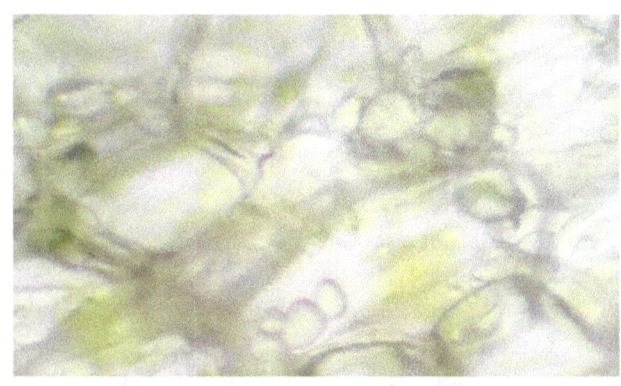

This is a thin slice of potato that is starting to go green. Some of the amyloplasts are making chlorophyll.

We've already learned what proteins are made of: amino acids. All three of the remaining foods on your plate contain long strings of amino acids that your digestive system will break apart into individual units so that your cells can use the amino acids to build their own cellular structures. Some of the amino acids you eat will be used to make the "pumps" and "motors" found in the membranes of your cells.

Your piece of steak is a slice of muscle and bone from a cow. You are not going to eat the bone, of course, so let's set it aside for now and let you study it in an anatomy course. The red part of the steak is made of muscle tissue. During the cooking process, the structure of the muscle tissue probably changed quite a bit, but let's pretend that it didn't change very much and find out what muscle is made of.

A muscle is made of bundles of bundles of bundles. The smallest bundle is called the **myofibril** and it is made of individual muscle fibers. Muscles don't have individual cells like other body parts do. When the muscle was forming, its cells fused together to make long fibers that have many nuclei. The fibers are packed full of two long proteins called **actin** and **myosin**. (In the diagram on the bottom, actin is blue and myosin is red.) Those tiny bumps on the red myosins are like little oars that can push against the actin and cause it to slide forward. When sliding happens to thousands of fibers simultaneously, the muscle contracts, causing a body part to move. Looking at this diagram, we can see why meat often has a directional, "stringy" texture.

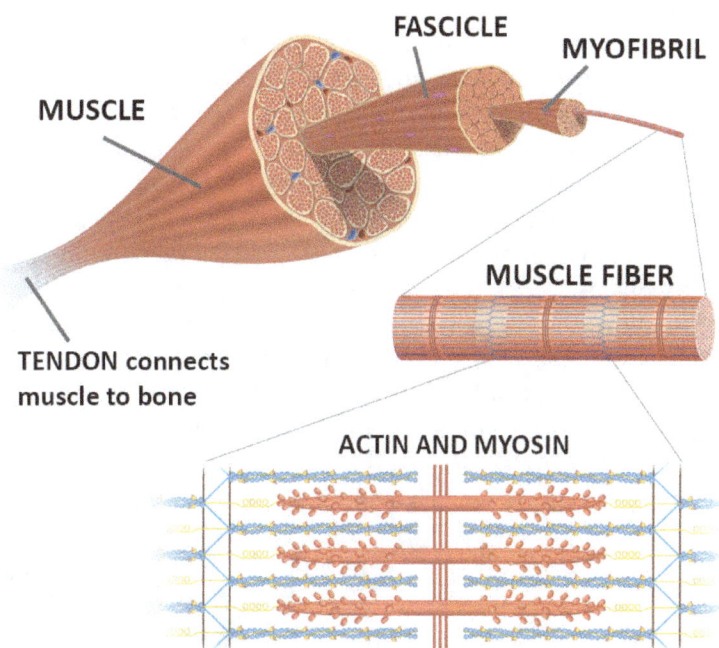

The digestive enzymes in our stomach and intestines will tear apart actin and myosin, breaking them into individual amino acids. Actin and myosin provide 18 of the 20 different types of amino acids.

Muscles are connected to bones by tendons. Tendons are also made of proteins (such as collagen), and they continue up the muscle, becoming more like a thin plastic bag as they cover all the bundles. If you want to observe tendons, the best place to see them is in chicken "drumsticks."

Many people think that red meat is red because it has blood in it, but fortunately this is not true. The red color comes from a molecule called **myoglobin**. Myoglobin is similar to **hemoglobin**, the molecule that picks up oxygen in our lungs and carries it through the blood. At the center of hemoglobin is a molecule of **heme**. We met heme in a previous chapter and learned that it is similar to chlorphyll. Both have a central "X" shape in the middle, made of four nitrogen atoms holding on to either a magnesium atom (chlorophyll), or an iron atom (heme). It is the iron atom in heme that causes it to reflect red light and thus make our blood appear red.

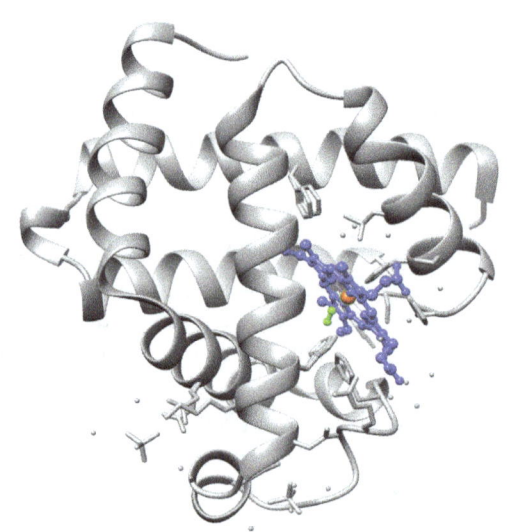

Myoglobin also has a heme at the center. In this diagram, heme is dark blue. The iron atom at the center is red, and the oxygen molecule (O_2) is green. The gray parts are proteins made of amino acids.

Myoglobin holds oxygen atoms until the muscle fibers need them. If you want to find out how long it takes for all the oxygen in your myoglobin to get used up, just hold your breath. If you stop taking in oxygen, your cells will begin to deplete all sources of stored oxygen. Animals that can hold their breath for a very long time, such as whales and seals, have a lot more myoglobin in their muscles than you do.

When the heme in myoglobin has an oxygen molecule stuck to it, it appears bright red. That's the red color in your meat. People like raw meat to be red, not brown, so meat packers expose the raw meat to oxygen, often in the form of carbon monoxide, CO, which sticks even tighter to heme than pure oxygen does. (This is why carbon monoxide is so dangerous!) The heme will hold on to that molecule of CO or O_2 for days, even a week or two if kept refrigerated. If meat is brown, that doesn't automatically mean that it has spoiled; it might have just lost its oxygen atoms. Smell is a better way to tell if meat is going bad.

Different animals have different amounts of myoglobin in their muscles. Fish have very little myoglobin, so their muscles don't look red. (Chickens have more myoglobin in their leg muscles, and less in their flight muscles, giving us "dark meat" and "white meat.") The orange color found in fish such as salmon comes from a carotenoid pigment called **astaxanthin** *(ast-ah-ZAN-thin)*. The fish get this pigment from the shrimp and krill that they eat. Flamingoes are orange for the same reason; they eat tiny orange animals.

The group of molecules that fish are most famous for are the **omega-3 fatty acids**. If we put our magnifier on its highest setting, we can zoom in on one of these molecules and find out why they are called "omega-3." The molecule you see in the viewer is a "space filling" model that puts the atoms right next to each other instead of using a line to show the bonds. (Black is carbon, white is hydrogen and red is oxygen.) The red end is where the molecule would attach to a glyercol "hanger" creating triglyderides or phospholipids.

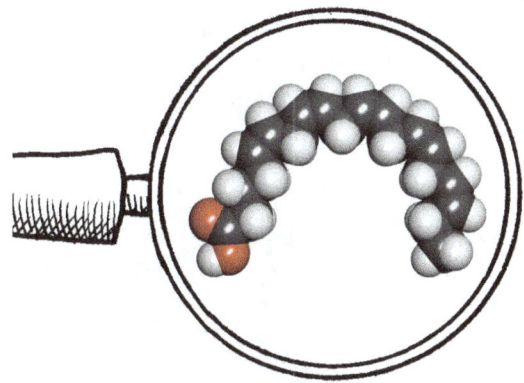

Omega is the last letter in the Greek alphabet, therefore this word is sometimes used to name things that are last in a line. Here, the line is made of carbon atoms, so the last carbon atom in the chain is the "omega carbon." The third carbon from the end is therefore the "omega-3 carbon." This is an omega-3 known as EPA: (hydrogens not shown)

The significance of the omega-3 carbon is that it is the first carbon that doesn't have two hydrogens attached to it. One hydrogen is missing, forcing that carbon atom to "double up" and form two bonds with the next carbon down the line. We show this using a double line. Every place you see a double line, there is a hydrogen atom missing. Carbon chains like this are called **unsaturated**. Chains that have all of their hydrogens are called **saturated**.

A carbon chain that has all its hydrogens is straight. Hydrogens don't like to be next to each other (their negative charges repel), but when the carbon chain is completely full of hydrogens (saturated), the hydrogens have no chioce but to sit there in a straight line and be unhappy. When a hydrogen leaves, this relaxes the situation and allows the hydrogens to shift their positions, and we see this as a slight curve. The molecule in our viewer has quite a few missing hydrogens, so it curves a lot.

So why are we told that omega-3 fatty acids are good for us? We don't know for sure, but one leading theory says that when these curved fatty acids are incorporated into our cell membranes as the tails of phospholipid molecules, the function of the membrane improves, especially in nerve cells. Some scientists speculate that the curved shape acts as a detangler as it moves around in the membrane. It is nearly impossible to observe a membrane "in real time" at the magnification necessary, so we might never get a definite answer. Also, some researchers say the claims about the health benefits are somewhat exaggerated. (Claimed benefits include lowering inflammation, lowering cholesterol, decreasing risk of cardiovascular disease, normalizing hormone levels, and improving skin conditions.)

Yes, we are ready to reveal the science inside that mysterious little patty! You are right—it looks, smells and tastes like a normal hamburger. You'd never guess that it wasn't real meat. It is made of "plant-based meat," which sounds like an oxymoron (a word or phrase that is contradictory) but this term will make sense once we explain it.

There have always been meat alternatives for people who don't eat meat, but the patty on your plate is something new. It is the result of high-tech research by PhD chemists and food scientists who wanted to find out exactly what chemicals give "red meat" (usually from cows) its characteristic smell and taste. What is it that meat-lovers love? They discovered that the key molecule is one we've already met: **heme**. Heme is what people crave. If they could find a source of heme that came from plants instead of animals, it might be possible to make a plant-based meat that tasted like real meat.

Several decades ago, plant scientists had discovered a heme-like molecule in tiny nodules (bumps) on the roots of soybean and alfalfa plants. The roots of these plants make these tiny nodules to house a certain species of bacteria found in the soil. These bacteria can do something that plants can't do—they can take nitrogen out of the air and "fix" it into a form that plants can absorb. The bacteria are often called "nitrogen fixing" bacteria. Before nitrogen fertilizers were invented, farmers had to plant soybeans or alfalfa in their fields every three or four years. Crops like wheat or corn would take most of the nitrogen out of the soil and only soybeans or alfalfa could put nitrogen back into the soil.

Soybean root nodules that host "nitrogen fixing" bacteria

When it became possible to analyze this heme-like plant protein, they discovered that it was most similar to myoglobin, though also very simimlar to hemoglobin. It was named **leghemoglobin**, using "leg-" from the word "legume," which is the name of the plant family to which soybeans belong.

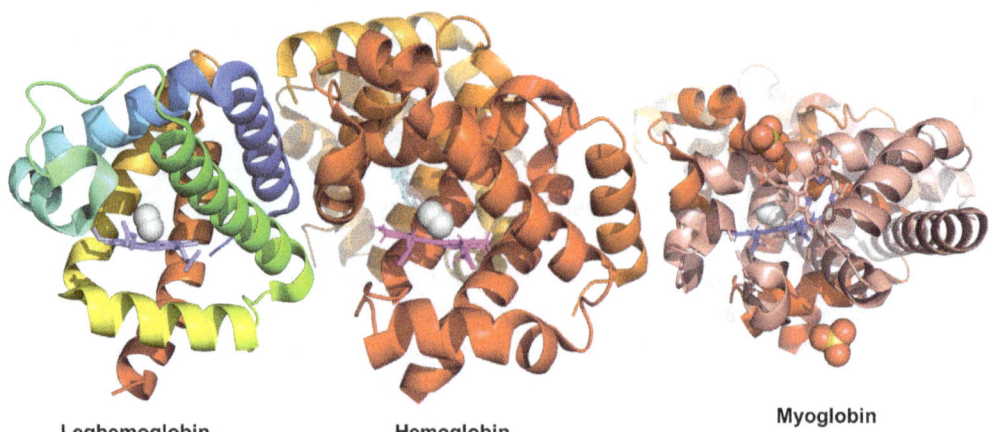

Leghemoglobin　　　　Hemoglobin　　　　Myoglobin

What a strange place to find heme! After more research, the scientists figured out why a myoglobin-like molecule was necessary inside these root nodules. The bacteria living in the nodules don't like oxygen. They need an environment with little or no oxygen. The mitochondria in the plant cells, however, need a plentiful supply of oxygen so they can make lots of energy molecules (ATPs). The leghemoglobin molecules provide a solution to this problem. They are able to grab and hold any stray oxygen molecules that come into the nodule, reducing "free" oxygen, which makes the bacteria happy. The leghemoglobins are also able to release the oxygen atoms into the plant cells, giving the mitochondria the oxygen they need for recharging ATP molecules.

The next problem was how to harvest enough leghemoglobin. Soybean root nodules take months to grow and contain only a small amount of leghemoglobin. They needed a way to scale up production so that a plant-based meat factory could produce tons of meat every day. The solution was to use genetic engineering. They were able to find the place in the soybean's DNA that contained the instructions for how to make leghemoglobin. They snipped it out and then inserted it into the DNA of a yeast cell. Now they had yeast cells that could make leghemoglobin! Yeast cells are very easy to grow in large quantities. Breweries have huge vats of them.

yeast cells

The meat researchers also found that there are specific amino acids and certain types of fat that combine in Maillard reactions to give that "frying beef" smell and flavor. They used super high-tech machines to analyze the aroma of cooking meat, to give them clues about what combination of plant products might give similar results. They came up with a combination of sunflower oil and coconut oil, soybeans, potato protein gel, and cellulose. Vitamins and minerals were added, also. They used professional food processing machines to chop and mash the ingredients until the texture was just right.

Some scientists are skeptical about whether this new type of plant-based meat will turn out to be safe for people to eat on a regular basis. The safety study submitted to the US FDA (Food and Drug Administration) in 2017 was a one-month study on rats. The rats were fed a much higher level of leghemoglobin than any person would ever eat, and then after a month the rats were humanely killed and all their body parts were examined.

The study found that the rats had experienced some negative changes to their body chemistry, though nothing so drastic that they look or acted sick. We'll never know what would have happened to the rats had they kept eating leghemoglobin for their entire life. It is possible that eating leghemoglobin meat carries the same risk that eating regular red meat does. Studies have shown that people who eat lots of red meat are more likely to have certain health issues as they grow older. Heme is a likely culprit for the root cause of these problems, but cooked meat (especially crispy or burnt meat) contains other types of mildly harmful molecules, too. It's hard to design a long-term food study that can isolate one ingredient.

This information isn't meant to scare anyone from eating red meat or trying the new leghemoglobin meat. With any food, we should eat in moderation.*

* Nothing in this book is intended to be medical advice about what you should or should not eat. The author is not for, or against, eating meat or meat substitutes.

Comprehension self-check

1) Capsaicin is a chemical found in this food: _____, that can trigger a cell membrane portal (in one of your tongue cells) that is normally triggered by: a) heat   b) cold   c) pain

2) Are animals sensitive to capsaicin?   Are birds?

3) Bean pods are what (reproductive) part of the bean plant?

4) From the plant's point of view, what are its cotyledons for?

5) Nitrogen and sulfur are found in: a) proteins   b) fats   c) sugars

6) Oil bodies (in seed cells) contain what "3-legged" molecule?

7) Starch granules are made of strings of _____ molecules.

8) Name two plants that have an endosperm and only one cotyledon in their seed:

9) What does "agglutinate" mean?

10) What does the heme molecule do in your body?

11) Phytohemagglutinin belongs to a family of molecules called l_____, which are proteins that bind to:
    a) other proteins   b) sugar strings   c) fatty acids   d) cell membranes

12) What can you do to neutralize (get rid of) the toxic chemical found in beans?

13) A short string made of three sugars that only bacteria can digest:

14) The names of sugars end in "__ __ __" and the names of enzymes end in "__ __ __."

15) Will cooking a green potato get rid of the solanine toxins?

16) Which grows better in cold climates, potatoes or sweet potatoes?

17) TRUE or FALSE? An amyloplast can turn into a chloroplast.

18) Which are stem tubers, potatoes or a sweet potatoes?

19) Where would you be more likely to meet a chromoplast, in a potato or a sweet potato?

20) TRUE or FALSE? Meat is red because it contains blood.

21) What two proteins slide against each other in a muscle fiber?

22) What molecule holds oxygen atoms in a muscle fiber?

23) The smallest "bundle" in a muscle is called a:

24) To make raw meat keep its red color, it is sometimes exposed to _____ _____ during packing.

25) Where did the salmon get its orange pigment molecules?

26) What are omega-3 fatty acids missing? a) oxygen atoms   b) carbon atoms   c) hydrogen atoms

27) The roots of soybean plants have tiny nodules that contain _____ that can take _____ out of the air and make it available to plants.

28) What organism makes the heme used in plant-based meat?

29) What reaction creates delicious flavor molecules? the M_____ reaction

30) Beans belong to the _____ family of plants.

ACTIVITY 5.1   Matching

Each word in the numbered list has some kind of association with a word in the letter list. The associations are of different types, but you should be able to figure out which words go together. Write the correct letter in each blank. (Use the process of elimination if necessary, and start with the ones that you are sure of.)

1) ___ beans
2) ___ potatoes
3) ___ amyloplasts
4) ___ peppers
5) ___ seed leaf
6) ___ true leaf
7) ___ corn
8) ___ raffinose
9) ___ sweet potato
10) ___ chloroplast
11) ___ myosin
12) ___ heme
13) ___ salmon
14) ___ triglyceride
15) ___ amine

a) actin
b) astaxanthin
c) cotyledon
d) capsaicin
e) endosperm
f) fatty acids
g) green

h) nitrogen
i) plumule
j) phytohemagglutinin
k) solanine
l) alpha-galactosidase
m) starch granules
n) root tuber
o) oxygen

ACTIVITY 5.2   What are these things?

Match a word to each picture. All the answers are given at the bottom...plus a few extra words you won't use.

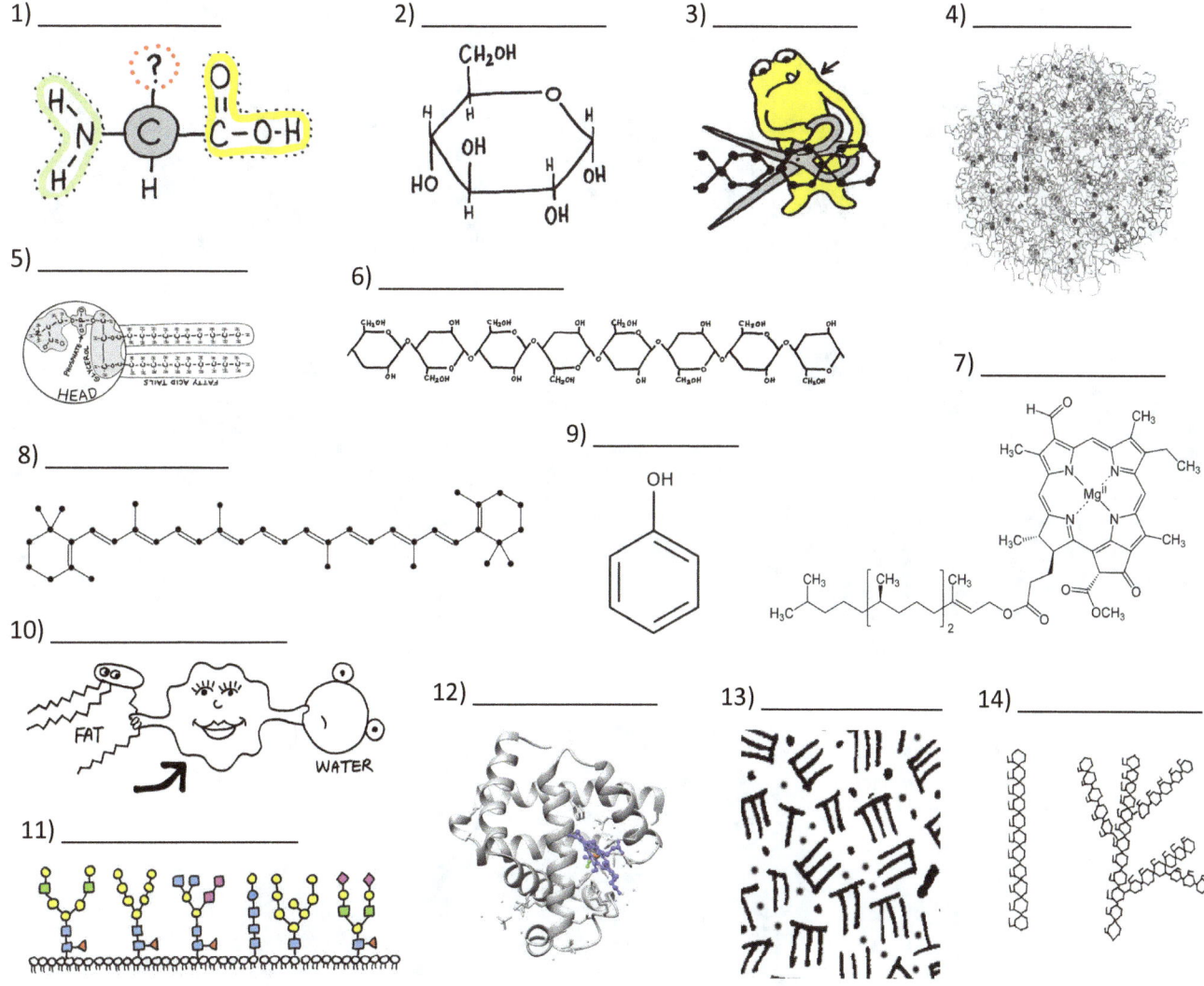

1) _____
2) _____
3) _____
4) _____
5) _____
6) _____
7) _____
8) _____
9) _____
10) _____
11) _____
12) _____
13) _____
14) _____

enzyme, phospholipid, glucose, cellulose, carotene, micelle, amino acid, phenol, oligosaccharides, myoglobin, myosin, chlorophyll, sucrose, triglycerides, emulsifier, starch, gluten

ACTIVITY 5.3   Meet the "nightshades" (Solanaceae family of plants) *(so-LAN-uh-SEE-uh)*

In this chapter we learned that potatoes and peppers and tomatoes all belong to the same family of plants, the Solanaceae, often called the "nightshades." All the members of this family make an **alkaloid** chemical. We've met two of these alkaloids: **capsaicin** and **solanine.** Another alkaloid you may have heard of is **nicotine**, the addictive chemical in cigarettes. Nicotine is made by tobacco plants, which are members of the Solanaceae.

Of the four major types of alkaloids, the **tropanes** are the least well-known. One of the most interesting and useful tropanes is **atropine**, a chemical that in high doses is a harmful stimulant, but in low doses becomes an essential medicine. Eye doctors use atropine eye drops to dilate pupils so they can get a better look at the inside of their patients' eyes. The action of atropine on the eye was discovered a long time ago. During the Renaissance period, Italian women would take the juice of the **"bella donna"** plant (also known as the deadly nightshade) and squeeze it into their eyes to open their pupils, thinking that this would make them more attractive. Another medical use for atropine is to reverse the effects of insecticide poisoning or exposure to gases used in chemical warfare.

One of the most unusual flowering plants is a member of this family: the **moonflower**. The flower is large and white, and has 5 petals that unfurl in the evening. The flower stays open all night then closes in the morning. The flowers smell good, but if you crush the leaves or stems they are said to stink like rotten peanut butter. Moonflower plants produce atropine as well as hyoscine, a chemical used to treat motion sickness.

There are thousands of plants in the Solanaceae family, many with strikingly beautiful flowers.

bella donna   jimson weed   petunia   moonflower unfurling   flower of potato

many varieties of tomatoes   tomatillo   many varieties of peppers   eggplant (aubergine)   tobacco

This is a famous painting by American artist Georgia O'Keefe.

The name of the painting is "Jimson Weed."

ACTIVITY 5.4   Fifth installment of "Chew It Over," a group game to be played during a meal

Here is another round of questions for you to use at a mealtime that you share with family or friends. These questions relate to the topics we learned about in this chapter. Again, you can use these questions in a varity of ways. You can be the quiz master and determine who gets which questions, or you can cut the questions out of the book and put them into a bag or bowl and let people choose a question randomly. The answers on are the back of this page.

CHAPTER 5: MAIN COURSE 1) Which U.S. state as a potato museum and is known as the potato state? a) Idaho b) Montana c) Oregon d) Washington	CHAPTER 5: MAIN COURSE 2) How many species of fish can you name that are likely to be in your local food store?
CHAPTER 5: MAIN COURSE 3) What are these animals called once they are cooked and served for dinner? (Name a meat for each one.) a) cow b) sheep c) pig d) young calf	CHAPTER 5: MAIN COURSE 4) JOKES Why is it so hard to keep a secret on a farm? What day of the week do potatoes hate?
CHAPTER 5: MAIN COURSE 5) There is a hotness scale for peppers, called the Scoville scale. Green bell peppers score less than 100. Jalepeños can score as high as 10,000. Guess the score of the hottest peppers in the world (such as Carolina Reaper).	CHAPTER 5: MAIN COURSE 6) Guess which country eats more potatoes--United States or Germany?
CHAPTER 5: MAIN COURSE 7) WHICH STATEMENT IS FALSE? 1) Marie Antoinette wore potato flowers in her hair. 2) Napoleon ate potatoes for breakfast every day. 3) Thomas Jefferson introduced French fries to America.	CHAPTER 5: MAIN COURSE 8) WHICH STATEMENT IS FALSE? 1) All peppers start out green, then turn yellow or red. 2) Birds can't sense the hotness of spicy peppers. 3) Peppers were discovered about 1,000 years ago. 4) Bell peppers can be purple.
CHAPTER 5: MAIN COURSE 9) How many varieties of potatoes can you name? How many ways of cooking pototoes can you name?	CHAPTER 5: MAIN COURSE 10) WHICH STATEMENT IS FALSE? 1) Peppers have more vittamin C than oranges. 2) Peppers originally came from India. 3) Sweet peppers and hot chili peppers are members of the same species.
CHAPTER 5: MAIN COURSE 11) Which person at your table can tolerate the hottest spices? Which person has the lowest tolerance for spicy foods?	CHAPTER 5: MAIN COURSE INTERESTING FACT: In November, 2021, a family in New Zealand accidentally discovered a giant potato growing in their garden. It weighed over 17 pounds (8 kg). They named it Doug. Last we heard, Doug was still in their freezer.

1) a) Idaho
2) Answers will vary. In the U.S., common species are tilapia, salmon, mahi mahi, tuna, sole, halibut, cod, perch, flounder.
3) cow: beef, steak / sheep: mutton / pig: pork (bacon, ham) / young calf: veal
4) Because the corn have ears, the potatoes have eyes, and the bean stalk. / Fry-day   5) 1.6 million
6) Germany. The average German eats over 200 pounds per year, the average American eats about 140 pounds.   7) #2
8) #3 Peppers have been discovered in archaeological sites dating back to 2,000-3,000 BC.
9) Popular in USA: Russet, Yukon gold, fingerlings, red, white, Adirondack blue, Kennebec (Your area may have others.)
Mashed, baked, French fries, hashbrowns, potato pancakes, twice-baked, gnocchi, au gratin, tater tots, potato salad, potato chips, stuffed potato skins, potato soup, and others you may have in your area.
10) #2 Peppers are native to South America.

6: Dessert

Lemon meringue *(mer-RANG)* pie gives us three things to study: the yellow filling, the fluffy, white meringue, and the pie crust. The filling and the meringue are both made from eggs, but one uses the yolks and the other uses the whites. Before we get into the chemistry of each, let's look at the inside of an egg.

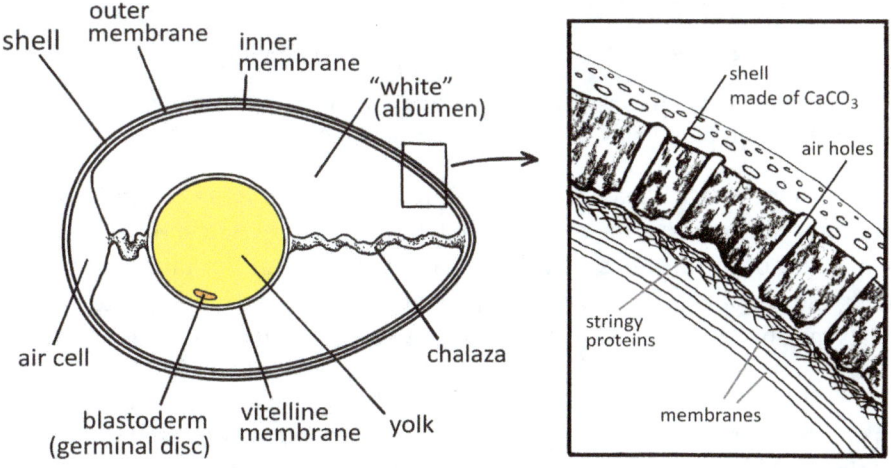

The **shell** is made of the same stuff as limestone rock—**calcium carbonate**, $CaCO_3$. There are two thin **membranes** just inside the shell; they serve as a barrier to bacteria, and they keep the eggs from drying out. The outer membrane sticks to the shell and the inner membrane sticks to the albumen.

Egg white, also called **albumen** (from the Latin "albus" meaning "white"), is 90% water and 10% protein. The protein in egg white is a mixture of proteins, not just one type, but for our purposes we don't need to know all the names of these proteins. The role of the egg white is to insulate and protect the yolk, and also to provide proteins for the developing bird embryo if the egg happens to be fertilized (with sperm by the male before the egg develops) and is kept warm by an adult bird, or inside an incubator device at a hatchery.

Egg yolk is full of all kinds of things—vitamins, minerals, proteins, fats, beta-carotene (which provides the yellow color), and **lecithin**, a type of emulsifier. (Remember how we used eggs as an emulsifier for making mayonnaise?) The reason that the yolk is so rich in nutrients is because it is the primary source of nutrition for a growing bird embryo. The part labeled **blastoderm** is where an embryo will start to grow in a fertilized egg. (Please note, however, that eggs you buy in a store can't ever grow into a baby bird. Hens will lay eggs regardless of whether there is a rooster around to fertilize the eggs. If an egg is not fertilized, it has only half the amount of DNA it needs and therefore can't grow an embryo. Even if you get your eggs from a farm that keeps roosters, the eggs will have been kept at a temperature that prevents the growth of an embryo.)

The **chalaza** is the anchor that keeps the yolk in the middle of the egg. The **air cell** is always at the large end of the egg and provides a good place for gas exchange, allowing oxygen to come in and carbon dioxide to flow out. Gas exchange can occur in other parts of the egg, also, because the shell has over 10,000 microscopic holes.

The process of making a lemon meringue pie starts with separating the eggs. You crack an egg and then very carefully tip the halves back and forth, allowing the white to slop out into a very clean bowl. You have to be careful not to let even the tiniest bit of yolk get into the whites because the chemistry of the fluffy meringue depends on it.

The proteins in egg white come in little clumps. This is their "natural" state. When egg whites are whipped, the little balls are disrupted and they come apart. After the proteins unfold, we say that they have been **denatured** (taken out of their natural state).

Whipping the egg whites also introduces lots of tiny air bubbles. (Cooks have discovered that the whites will whip better at room temperature than at refrigerator temperature. Cold proteins apparently don't unfold as well as warm ones.) The denatured proteins are attracted to the air bubbles and begin to stick around the edges. Soon the air bubbles have been completely coated with a thin layer of proteins. We might imagine these bubbles as tiny balloons made of proteins instead of latex rubber. The end result is a massive network of microscopic protein balloons all stuck together. These little balloons can be made more stable by cooking the meringue. This toughens the proteins and helps the balloons not to pop.

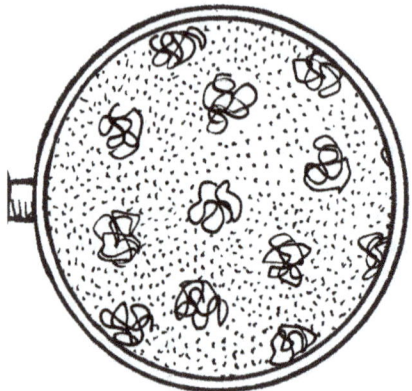 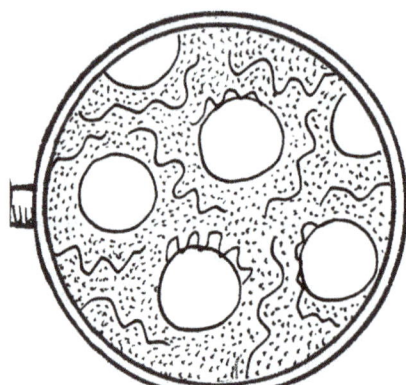 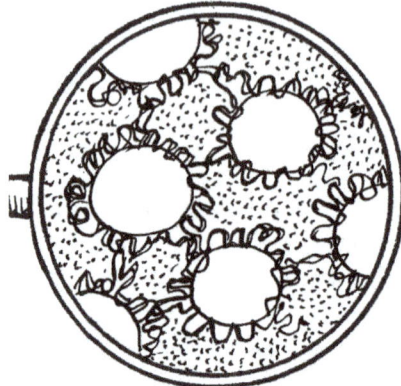

1) Egg white before whipping, with clumps of proteins.

2) Egg white during whipping. Proteins unfold and air bubbles appear.

3) Proteins stick to the edges of the air bubbles and stabilize them.

Another trick that cooks use to stabilize the meringue "sponge" is to add some cream of tartar and some sugar during the process of whipping the egg whites. However, these substances slow down the denaturing of the proteins, so you can't add the sugar and/or tartar too fast. You add a little at a time while doing the whipping. The key is to be patient and allow the meringue to develop slowly. When the meringue starts making tall peaks with pointed tops, you know it is time to turn off the beater.

Before modern times, when cooks did not have access to all the conveniences we have today, such as baking powder and gelatins, eggs were the primary ingredient used to thicken liquids and to help baked goods attain a lighter texture. The chemistry behind these uses for eggs is essentially the same as the process we just saw in the meringue. Heat can also cause the egg proteins to denature and then rearrange themselves.

The egg yolks in our filling start out looking a lot like the whites, but with yellow beta-carotene.

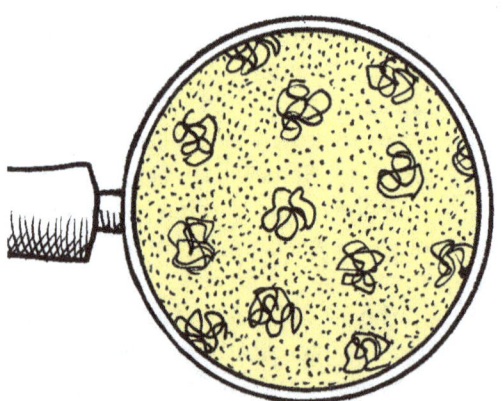 

All the fats, lipoproteins and emulsifiers in egg yolks slow down the processes of denaturing and coagulation (reforming into new structures). This is the reason that egg custards take so long to bake in the oven. Adding milk or water can slow down the process even more, but the result is an enjoyable, smooth, mildly firm texture.

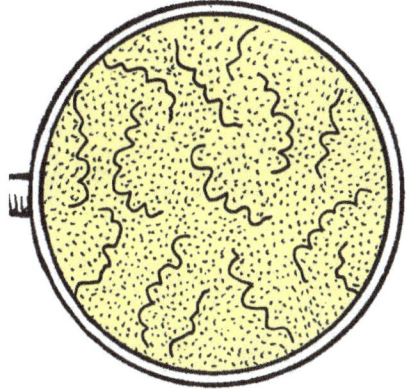
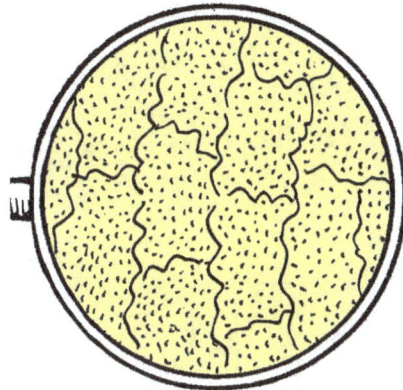
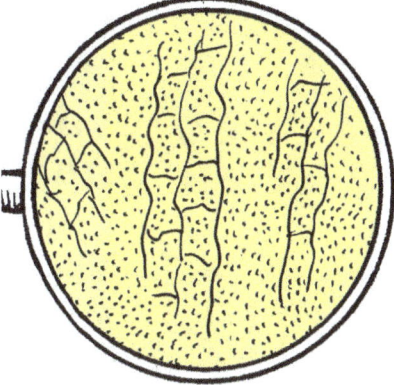

1) Heat causes the clumped proteins to unfold and become "denatured." Water flows freely around them.

2) The proteins begin to coagulate and form new structures. Water is trapped in pockets and cannot flow.

3) This is what happens when egg proteins are over-cooked. The structures shrink and water is released.

What would the textures of the above pictures be like? In the first picture we have water and proteins floating around freely, so the mixture would be watery in our pan. In the second picture, protein structures have created "pockets" trapping water molecules inside so that water can't flow freely. This texture would be a soft solid, harder than pudding but softer than jello. In the third picture, we left the egg custard in the oven a little too long and the protein structures shrank and became tighter. If we took a bite of this, we'd find it to be very runny, with rubbery strings or clumps of protein. We should have set a timer!

The recipe the chef used for the lemon filling also called for some cornstarch, to add additional thickening. (If we had used whole eggs, we'd have had extra thickening help from the whites.) We saw starch granules in the last chapter. They are inside amyloplasts and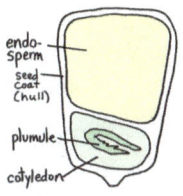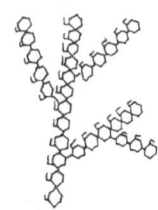
are made of long strings of glucose molecules, either in straight lines or in branched shapes. In corn, the starch granules are found in the endosperm part of the seed.

The cornstarch is combined with some water and sugar and then put on medium heat and stirred constantly so the starch won't stick to the bottom or sides of the pan. As the solution warms up, the starch granules start to absorb water. (Imagine balloons slowly filling with water.) As the granules swell, the solution starts to get a little thick. Then, suddenly, many of the granules burst and the amylose chains spill out. The chains get all tangled up and hold water in little pockets so it can't flow freely (as we saw with the eggs).

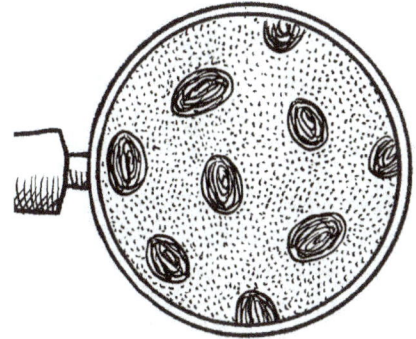
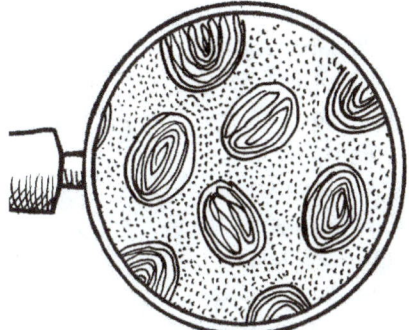
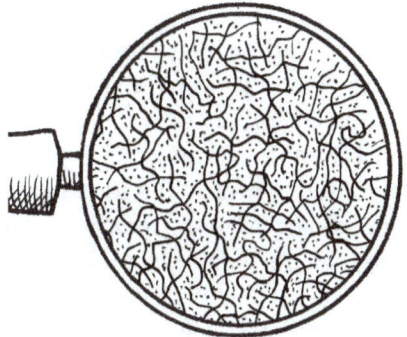

1) Starch granules floating in water. The granules contain amylose chains.

2) Starch granules swell up with water as the water becomes very warm.

3) Starch granules explode and release amylose. Water can't flow very well.

Since this is not a cookbook, we won't go into all the details about exactly how to make lemon meringue pie. We just need to make sure we've looked at the molecules. We just learned about eggs and cornstarch, and in past chapters we've discussed sugar, salt, water, cream of tartar, butter and vanilla. The remaining essential ingredient—lemon juice—has chemicals that belong to categories we've already discussed: vitamin C, B vitamins, flavonoids, natural oils, acids, and amino acids. So let's move on to looking at that pie crust.

Some people make pie crust with just butter, but many people use a combination of butter and shortening. We decided to put shortening into our crust so that we could yak on at you about the chemistry of hydrogenated oils. You don't have to eat the crust, but you do have to learn about its molecules. (If you can't remember a thing about butter, you might want to take a few minutes to go back and review those pages at the beginning of chapter 3. We're going to dive right into shortening chemistry.)

Shortening is usually made from soybean oil. We saw "oil bodies" in bean seeds on page 83. (If you would like to see the process of oil extraction, there are a few videos posted on the youtube channel.) All vegetable oils are made of those 3-legged triglyceride molecules. The "legs" are made of chains of carbon atoms that are usually about 16 to 18 carbons long. Most of these carbon chains have at least one double bond, and many of the chains will have several double bonds. A double bond means that there are two hydrogen atoms "missing" and the carbon atoms at that place are having to improvise by "holding hands" with each other instead of with hydrogens.

soybean pods

harvested beans

Carbon atoms make four bonds. We can imagine these bonds as "arms" that are holding on to other atoms. In our illustration, the lines are the "arms." Notice that each C has four lines going out from it, though the lines are often shared. In the saturated fatty acids, the carbons are holding on to the maximum number of hydrogens that they can possible have. In the mono-unsaturated fatty acid, we see one place ("mono" means "one") where two H's are missing. The carbon atoms do a double handshake to compensate. Unsaturated means "not completely full."

Some of the fatty acids we find in soybean oil are poly-unsaturated, meaning that there are multiple places with a double bond due to "missing" hydrogen atoms. In a previous chapter we commented that when hydrogens are missing, the molecule will bend, due to the fact that the hydrogen atoms really don't like to be next to each other, so when a gap opens up they will shift their positions a bit. In the last chapter we learned that when the first double bond occurs at the third carbon from the end of the string, we call it an "omega-3" fatty acid. This unsaturated molecule is an omega-5, not an omega-3.

POLY-UNSATURATED FATTY ACID

Let's imagine our fatty acids to be plastic straws. Saturated fats would be straight straws, and unsaturated fats would be curved straws. Which type of straw would take up more space?

The straight straws line up very nicely into a compact bundle. The curved straws have an awkward shape and therefore can't be compact. We are told that saturated fatty acids, being a straight line, are like the straight straws and form a compact solid, like the white fat we find on red meat. Unsaturated fats are like the curved straws and are more loosely held together and therefore tend to be liquids, like olive oil or corn oil. Some natural oils, like coconut oil, are a mix of saturated and unsaturated, so they have properties of both. On cold days, your jar of coconut oil will be a white solid. On very hot summer days, your coconut oil will turn into a thick liquid.

The triglycerides in soybean oil are made of (mostly) unsaturated fatty acids. When the soybeans are pressed, the oil comes out as a liquid. If we can force these unsaturated molecules to accept hydrogen atoms, they will be turned into saturated molecules and therefore take on a more solid texture. This process is called **hydrogenation**. The oils are exposed to hydrogen gas in the presence of a catalyst, usually a metal such as nickel or platinum. The only role the catalyst plays is to provide an environment that encourages the carbon chains to open up and accept more hydrogens.

HYDROGENATION TURNS CURVED, UNSATURATED
FATS INTO STRAIGHT, SATURATED FATS

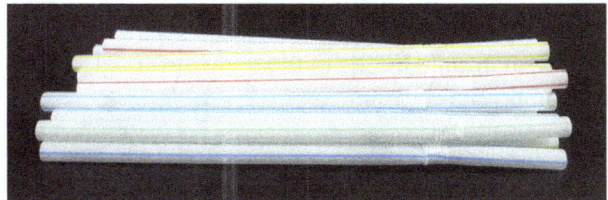

WHAT HAPPENS DURING HYDROGENATION:

Double bond opens to accept 2 H's.

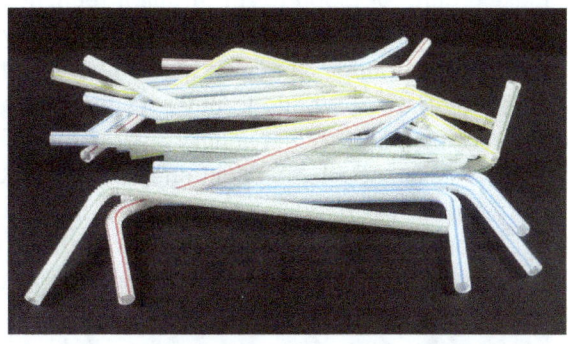

The process isn't perfect and some chains just can't be convinced that it is a good idea to open up their double bonds. When only some of the molecules have been changed, we saw that the oil has been **partially hydrogenated**. If you start reading labels on boxes of crackers and cookies you will often see "partially hydrogenated soybean oil" listed as an ingredient. Partially hydrogenated oils have an ideal texture for making baked goods.

Hydrogenation offers an additional benefit. Baked goods made with partially hydrogenated oils are more "shelf stable," meaning the product won't go stale as quickly. This is a huge advantage for large manufacturers who don't know how long their products might sit on a shelf at a small country store. They don't want unhappy customers complaining about boxes of stale biscuits. (Whoa, that's one unhappy customer!)

Hydrogenated oils seemed almost too good to be true. Modern science had made it possible to take inexpensive soybeans and turn them into a solid fat that had a perfect texture for baking AND had a long shelf life. All was well in the industrial baking world until someone discovered a seemingly minor issue that was having seemingly major effects on people's health.

There are two ways that these double bonds can be configured. One option is that the H's are on the same side of the bond. We call this the "*cis*" *(sis)* format. The other option is that they are on opposite sides. We call this the "*trans*" format. You've probably heard about "trans fat." In the past few decades there has been a considerable amount of research on these types of fat, and evidence has been growing to support the idea that cis fats are better for you than trans fats. It seems strange that such a small difference in a molecule can be so important.

What's so bad about trans fats? Scientists speculate that the trans shape tends to get stuck in bad places, such as along the insides of our blood vessels. A build-up of fat in small blood vessels will narrow their diameter and make it hard for blood to get through. Yeah, not good.

When researchers examined the partially hydrogenated oils that food companies were using, they found a considerable number of fatty acids with double bonds in the "trans" configuration. When this news reached the general public, many people went into a panic and some got very angry at large manufacturing companies. Governments began to ban all products that contained trans fats. Companies that processed vegetable oil had to quickly figure out a way to reduce the number of "trans" molecules in their hydrogenated oils. They seem to have succeeded, and we now see "NO TRANS FAT" proudly printed on the labels of many food products.

Here is the list of ingredients in our shortening: **Soybean oil, fully hydrogenated palm oil, palm oil, mono and diglycerides, TBHQ and citric acid.** Notice that "partially hydrogenated oil" is not in the list. Palm oil contains a mix of saturated and unsaturated fatty acids, but, because it is a natural product, is unlikely to have many trans fats. The last two ingredients, TBHQ and citric acid are preservatives that keep the oil from going rancid. (TBHQ stands for "tertiary butylhydroquinone.") We briefly discussed preservatives in chapter 2 and explained why food companies use them. If a company produces products that go bad easily, they don't stay in business long. People can say they don't want "chemicals" in their food, but they would be even more unhappy if their food was stale, rancid, or had bacteria growing in it.

Although this book is not about cooking, let's take a minute to discuss how pie crusts are made because they can show us how the *physical properties* of a substance can be manipulated. As a general rule, we don't have any chemical reactions going on here. All the ingredients keep their chemical identity.

The chef who made your pie crust had a tricky job. The flour and the fats had to be combined in a way that would make the crust both flaky and tender. The goal is to have thin layers of fat and flour that are not completely mixed together. Pie crust is the opposite of bread—you don't want to knead it. The less you touch it, the better. You roll it just enough to flatten it, and then stop.

A close-up cross section of pie crust would look like flat pieces of fat interspersed with flat bits of flour.

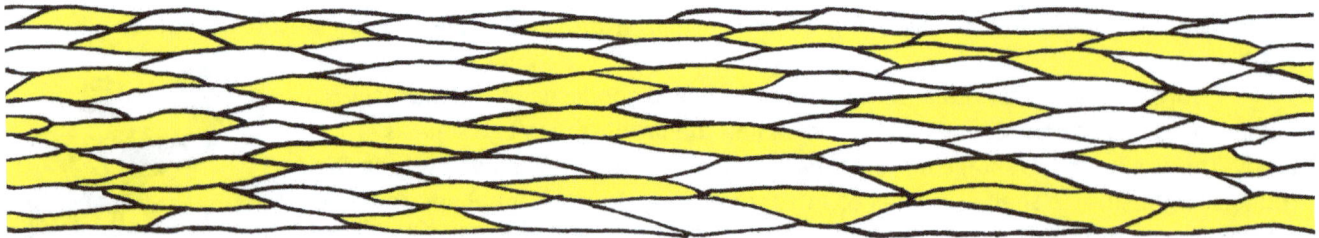

YELLOW IS FAT, WHITE IS FLOUR

The gluten in the flour is both friend and foe. Too little gluten and the dough will be impossible to roll. Over-developed gluten will make the crust tough. For this reason, "pastry flour" is lower in gluten than bread flour. When water is added to the dough (last step), the gluten problem will increase. Often, ice water is used because very cold water will be even less inclined to react chemically with either the flour or the fat. Other ingredients are often kept cold, as well. Some recipes recommend putting the dough into the refrigerator for a few hours before starting to roll it out.

Fortunately, the chemistry in this dessert will seem very familiar, because both the gelatin and the whipped cream involve thickening—a topic we've already been learning about.

The history of gelatin goes back to the 10th century (the 900s). In an Arabic cookbook that dates to this time period, we find a recipe for fish "aspic." **Aspics** are savory (not sweet) gelatins that contain pieces of meat, fish, vegetables or egg. The Arabic recipe called for boiling several fish heads with vinegar, parsley, onions, black pepper, and various other spices. Saffron was then added to give it a radiant red color. The fish heads were eventually removed but the eyes and lips were returned to the broth which was left to cool until it hardened.

By the 1300s, aspic technology had moved into Europe and French cooks found additional methods of gelling meat broths. In the 1400s, people in Britain discovered how to extract gelatin from boiled cow hooves.

In the 1700s, a French chef received a patent for a process that used cow bones to make gelatin. Soon after, it was discovered that the gelatin could be poured out into sheets and allowed to cool, making a product that could be stored and used at a later time. Gelatin continued to be sold in sheets until the mid 1800s when Americans discovered how to dry the gelatin and crush it into a powder. The brand name Jell-O® began in 1897 in New York. Jell-O® wasn't an overnight success. It took a decade of marketing and advertising before Jell-O® achieved the popularity we associate with it today. One advertisement from 1910 read like this:

Pork jelly is an aspic that is very popular today in Eastern European countries.

"Who likes Jell-O? The children. Do you remember the dreadful disappointment it used to be in the old days at home when mother brought on for dessert some baked apples or pie-plant pie, or something else that was "common," and you wanted shortcake or pudding? Now, the little folks want Jell-O. Every child loves Jell-O which is so delicious and refreshing, so full of nutriment, so pure and wholesome, so economical and so easily prepared, that there is no reason why the little tots, or anybody else, should be disappointed in dessert. The whole family like it just as well as the littlest members. A Jell-O dessert costs ten cents and can be made in a minute. It sounds almost too good to be true, but it isn't."

Jell-O was not immediately successful. It took years of advertising to make it into a heavily consumed product. The company gave away free cookbooks that had recipes for gelatin-based desserts.

Now for the chemistry of gelatin. We learned from our history lesson that this mysterious gelling agent could be extracted from fish heads, cow hooves, or cow bones. During the 1800s, the most popular source was pig skins, which the butchers would otherwise throw out. It wasn't until the 1900s that scientists really figured out what gelatin was made of.

The main ingredient is a protein called **collagen**. Like all proteins, collagen is made of strings of amino acids. The amino acid glycine is very prevalent in collagen because it is so small. It allows the strands to wind tightly together. Three strings wind together to create a single strand of collagen. Multiple strands of collagen wind together to form a bundle called a fiber. Collagen fibers are found throughout your body, and there are over 20 different types of collagen. Some collagens form a scaffold in your bones into which hard minerals can be deposited. Your skin is loaded with collagen types 1, 3 and 4, which make the skin both strong and stretchy. Collagen is the fabric that holds all your blood vessels and organs together. It's also found in the connective tissues in your joints.

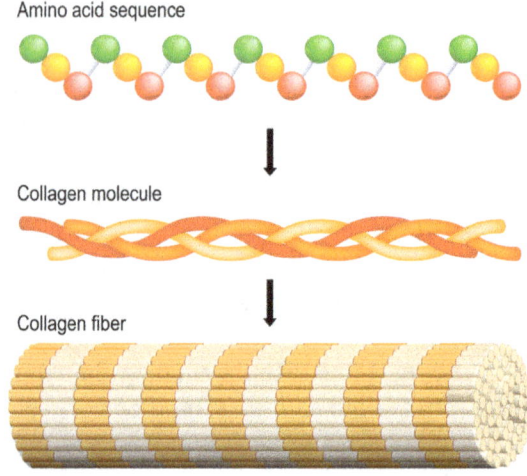

The diagram on the left explains why chefs have been able to extract gelatin from so many different animal body parts. Skin has the highest concentration of collagen, which is why pork skins took over the gelatin industry for many decades. Recently, there have been more options for consumers, such as kosher and halal gelatins made from fish or cow bones, or plant-based gelatins for vegans which are made from seaweed.

Extracted collagen has many other uses, besides making desserts. Many medicines come in clear capsules made of gelatin. Doctors use collagen products to heal wounds, and for reconstructive and cosmetic surgery. As long as a patient is not allergic to it, collagen from cows can be used in all these medical applications.

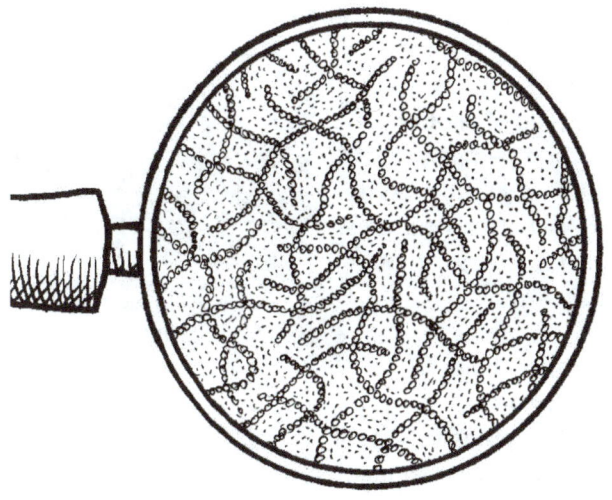

Now let's take a look at our gelatin dessert through our magnifier. We find a structure that is surprisingly similar to the lemon pudding that was thickened with eggs. We see long strands of protein (in this case collagen proteins) all tangled up so that the water can't flow around very well.

Water is the smallest molecule we see, but there are a few others almost as small, such as flavor molecules and pigment molecules. Our chef decided to use natural flavors and colors, so they would all be classified in one of those categories on page 67. Most instant jello mixes contain artificial dyes made from hydrocarbons that originally came from petroleum. We saw the Yellow #5 molecule back in chapter 2.

"Thickening" is a major theme in the world of desserts. In the whipped cream on top of the gelatin we see yet another example of something that was prepared by a thickening process. We were introduced to cream back in chapters 2 and 3, when we learned about milk and butter. Cream is made of water, a small amount of protein, and a lot of fat. The fat in cream is made of triglycerides that are inside globules surrounded by membranes. The cream is taken out of the milk and is used to make butter, or is sold as heavy or light cream, depending on the fat content. Only cream with at least 35% fat content will make whipped cream. Light cream has about 20% fat, so don't ever bother trying to whip it. Heavy cream has 38% cream, so it will make the best whipped cream. (Just for comparison, whole milk is 3-4% fat, and half-and-half is about 12%.)

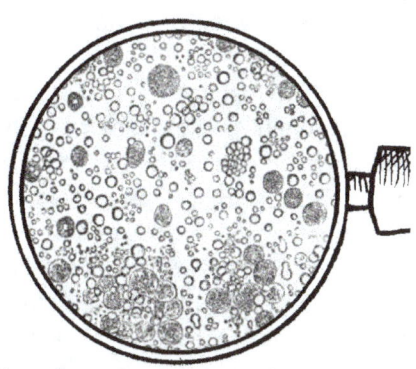

globules of fat in fresh "raw" milk

Whipped cream reminds us of meringue, except that here we have tryglcerides stabilizing the air bubbles, instead of proteins.

During the whipping process, the globules are broken apart and the triglycerides are released. If you think this sounds familiar, you are right. This is also the first step in making butter. In fact, if you whip your cream too long, you'll end up with butter.

Once the triglycerides are no longer inside their globules, they begin freaking out because they are floating in water. Fats hate water! They see a nearby air bubble and turn their tails towards the inside of the bubble. More and more triglycerides do this, until the air bubble is covered with them. The small proteins also feel a bit lost in the mess and will tend to associate with the outside of these fat-covered bubbles, adding an extra layer of stability. Floating between these stabilized air bubbles are molecules of water, sugar and vanilla flavoring. We saw the sucrose molecule in chapter 1, and we looked at vanillin in chapter 2.

Instant whipped cream is cream that has been dissolved into compressed nitrous oxide gas. N_2O. As the gas is released this process happens very quickly.

SUCROSE — PROTEINS — WATER MOLECULES — VANILLIN

Before we leave the topic of thickening, we should mention a few other substances that can do this.

AGAR: Made from red seaweed that has been boiled so that cellulose and other polysaccharides are released. The reasons agar might be used instead of another thickener are: 1) it is 100% vegetarian, and 2) it remains a gel at room temperature, unlike gelatin which must be kept refrigerated. Agar is more commonly used in Asian cuisine, and it is also used in biology labs as a "growth medium" for bacteria cultures. Nutrients required by the bacteria are mixed into the agar. The petri dishes can sit at room temperature or even be kept warm and the agar will remain firm.

Japanese dessert made from agar
By Sjschen - Own work, CC BY-SA 4.0, https://commons.wikimedia.org/w/index.php?curid=3068956

PECTIN: This is a natural substance found in fruit. The long molecular chains found in pectin are made of individual units called galacturonic acid (shown here). This is a galactose molecule with a COOH (acid) as the "flag" at the top of the molecule. Pectin is used to make jam and jellies. Jams that contain the whole fruit might have enough pectin in them to be able to "gel," but more often, extra pectin is added. Pectin is sold as a white powder, with the most well-known brand being Sure-Jell®.

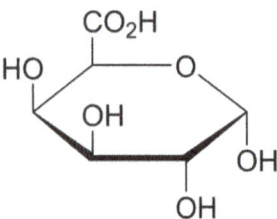

CARRAGEENAN: (called "E407" in Europe) This is another seaweed product. The word "carrageenan" comes from the Irish word "carraigin" which means "little rock." The red seaweed plant can be found growing on rocks along the coast of Ireland (where it is called **Irish moss**), as well along many of the rocky shorelines in northern Europe and North America. Seaweed isn't a plant, it is a type of algae. The carrageenan molecules that are extracted are "sulfated polysaccharides," which means "long chains of sugar molecules that have sulfur groups like SO_3 attached." The addition of sulfur groups tends to make the molecule more hydrophilic (water loving). Their hydrophilic nature makes them compatible with many milk and meat products, so carrageenans are used extensively in the manufacturing of ice cream, yogurt and other dairy desserts. It is natural product and can be included in products that are advertised as having "no artificial ingredients." Carrageenans are used in many non-food applications such as toothpaste, fire-fighting foams in fire extinguishers, shampoo, shoe polish, gel capsules for medicines, and in biotech they are used to hold cells and enzymes in place.

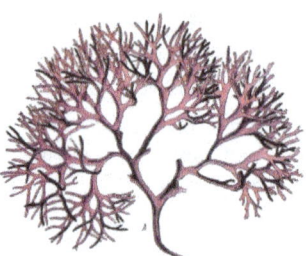

WHEAT STARCH: White wheat flour is the most popular choice of home cooks to thicken meat gravies. We know what starch is (very long chains of glucose molecules) so we don't need to say a lot more about it.

ROOT STARCHES: Another place we can find starch molecules is in certain types of roots. Tapioca starch and arrowroot starch are the best examples. These starches are soft, white powders and are especially good options for people who are allergic to corn and wheat. Tapioca starch comes from the roots of the cassava (manioc) plant. The word "tapioca" comes from the Tupi natives who lived in Brazil when the Portuguese explorers arrived in the 1500s. The word means "coagulated." No surprise there. Arrowroot plants grow in the tropics and the main producers today are the islands of St. Vincent and the Grenadines.

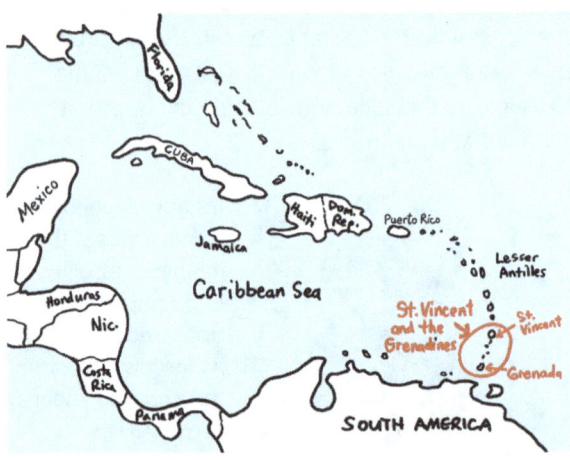

CASSAVA ROOTS

ARROWROOT PLANT
By Wibowo Djatmiko (Wie146) - Own work, CC BY-SA 3.0, https://commons.wikimedia.org/w/index.php?curid=10061180

Yes, there is a chocolate molecule! Well... sort of. The molecule that gives us the distinct chocolate flavor is **theobromine**. ("Theo" comes from the Greek word for "gods," and "bromine" comes from the Greek word for "food." Thus, we have "food of the gods," though the gods in this case weren't Greek, they were the Mayans of ancient Mexico.)

The plant that produces this chemical has the scientific name *Theobroma cacao*. The word **cacao** *(kah-cow)* is a very old Mayan word. We know how this word was written in the Mayan language. They used simple pictures, not letters, to represent syllables, so this word won't look like a word to you. Each part of the symbol stands for a syllable: "ka-ka-wa."

In the foreground we see an opened seed pod containing the raw seeds. In the background you can see the seed pods growing on the trunk of the tree.

The cacao tree grows seed pods on the sides of its trunk. When you cut open a pod, it looks nothing like chocolate. You see a bunch of white, slimy seeds. It's hard to believe that these seeds will become dark brown cocoa powder. The beans will be put into large wooden boxes where they ferment for almost a week. (You'll remember that we learned about fermentation in chapter 3.) The fermentation of cocoa beans is accomplished by friendly bacteria. After this, the beans are dried for several weeks. Once dry, the beans can either be shipped to other parts of the world, or they can be used locally to make chocolate products.

The next step is to roast the beans. Roasting changes the flavor and it also dries and loosens the thin hull surrounding the seed so that it can be removed easily. The roasted, hulled bean is then chopped into small pieces, called "nibs." The nibs are then ready to be ground into a fine powder. The powder is bitter and will need to be combined with a sweetener.

Half of the weight of cacao beans (or you can call them cocoa beans) is fat. If this fat is extracted we call it cocoa butter. It is a highly saturated fat and therefore has a long shelf life. Because it is so stable, it is used in beauty products, such as skin creams, that won't ever be refrigerated. 10 to 15 percent of the weight of the cacao bean is protein and the remaining 35 percent is carbohydrates, minerals such as potassium and magnesium, and a selection of chemicals related to theobromine.

Let's play "Spot the Difference" and look at four molecules found in cacao beans. Compare each one to caffeine (which we saw in chapter 2) and find what is different.

CAFFEINE   **THEOBROMINE**   **THEOPHYLLINE**   **PARAXANTHINE**

Caffeine seems to be the "parent" molecule from which the others are made. One evidence of this comes from physiology labs, where scientists have been able to track what happens to caffeine molecules after someone eats them. They discovered that when you consume caffeine, your liver will turn it into one of those other molecules as a first step in the digestion process. Interestingly, theophylline *(thee-OFF-ill-in)* is a molecule that is used as a medicine; it is given to people with asthma or similar breathing problems, as it causes the muscles in the airways to relax and open. Paraxanthine *(para-ZAN-theen)* has similar properties to caffeine, but isn't as potent; it is used in research but has no uses beyond that.

In addition to these chemicals, cacao contains many phytochemicals (flavonoids and phenols) that are antioxidants, helping to protect our cells from damage. This is why chocolate is often refereed to as a "superfood." However, it's not a superfood for our pets. Theobromine is manufactured by the cacao plant as a poison to kill bugs, and it can also have a toxic effect on dogs and cats. (Though the author's border collie once ate five pounds of marshmallow-filled chocolate candies and suffered no ill effects.) The biggest downside for humans of eating chocolate isn't the chocolate itself—it's the sugar that it is mixed with.

The other ingredients in your chocolate cake are things we've already discussed: flour (starch and gluten), sugar (sucrose), baking powder (baking soda plus a powdered acid), eggs, milk, and vegetable oil. Now let's look at what is on top of your cake. There's a dusting of powdered sugar, a drizzle of caramel sauce, and an edible flower.

Powdered sugar, also called confectioners' sugar, is table sugar (sucrose) that has been ground with an industrial blender until it is a super-fine powder. Usually, a little bit of cornstarch, potato starch, or tricalcium phosphate is added as an "anti-caking agent." These additives absorb moisture and keep the powder dry and fluffy. Two other types of powdered sugar that professional bakers use are "caster sugar," which is not as fine as regular powdered sugar and does not have any starches added, and "snow powder," a mix of glucose and starch that is resistant to dissolving and can be used on damp products like fruit tarts or even ice cream.

caramel apples

Caramel is plain old sugar (sucrose) that has undergone a chemical change. You'll remember that we saw some very odd chemistry happening in those **Maillard** reactions (chapter 3). Heat caused the breakdown of proteins, sugars and fats, and those broken molecules began sticking to each other to create "nonsense" molecules. We likened these "mutant" Maillard molecules to useless items created by merging appliances, tools, and toys. Those silly hodge-podge items certainly were strange. The reactions that happen in *caramelization* are similar to the Maillard reactions, but here we have only sugars, no proteins or fats, so our funny picture analogy won't work.

You can make caramel simply by heating sucrose, but usually a little water is added to slow down the process. Too much heat too fast will cause too much chemical activity and will result in molecules that taste bitter. Watching the temperature of the solution is a key to success, so cooks often use a "candy thermometer" when making caramel. The temperature of the solution must be kept under 375° F (190° C). You stir at first, but then, unlike many other recipes, you stop stirring and let the liquid boil undisturbed. Gradually, the liquid turns brown and you start to smell the new chemicals that are being formed.

In previous chapters, we saw sucrose and fructose separated by an enzyme. In caramelization, heat does the separation.

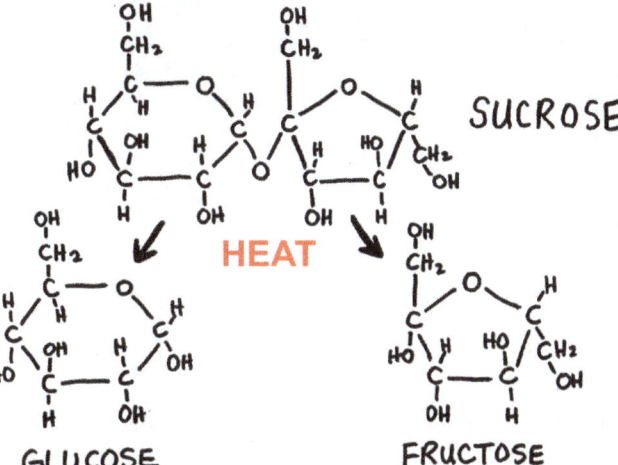

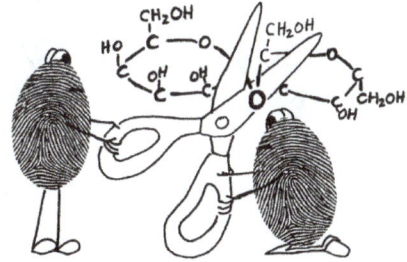

Remember this picture from chapter 1?

SIDE NOTE: A sugar syrup that is made of glucose and fructose molecules is called *invert syrup*.

After fructose and glucose are formed, the heat continues its work and breaks them apart, too. Broken pieces of fructose and sucrose begin bumping into each other and forming new molecules. Some of the new molecules will be small enough that they will go up into the air and eventually into your nose where you will smell them. Taken all together, they will smell like caramel, but here is how people describe the individual smells:

Diacetyl: "buttery"   Ethyl acetate: "fruity"   Furan: "nutty"   Maltol: "toasty"

Can you see where these molecules might have come from? (Furan and maltol are the most obvious.)

Some really large molecule are formed, too. You can't find pictures of these molecules; they are known by their formulas and their names: Caramelan: $C_{12}H_{12}O_9$ Caramelen: $C_{24}H_{26}O_{13}$ Caramelin: $C_{36}H_{18}O_{24}$
They are what makes caramel thicken.

The edible flower on top of your cake is made of **fondant**, a flexible blend of gelatin and sugar that can be rolled, cut, and molded into just about any shape. Fondant allows bakers to be artists. All the figures on this cake are made of fondant. You can eat the elephant!

Time and temperature play a huge role in sugar chemistry. As we mentioned during the caramel discussion, the texture, color, and taste of sugar can change as it is heated. Candy thermometers are marked with words like these: soft ball, hard ball, soft crack, hard crack. These words describe what the mixture in your pot would be like if you took it out as soon as that temperature was reached.

Fondant can be molded into any shape.

In hard candy that goes "crunch" when you bite it, the sugar has been cooked to the highest temperature on a candy thermometer, the "hard crack" (or "hard tack") level. It is a thick liquid when it is removed from the stove but quickly becomes hard as it cools.

Besides the sugar in your peppermint, there are quite a few flavor molecules. Peppermint plants produce over a dozen flavor-related molecules, but we're going to look at just a few. The molecule that makes peppermint feel cool in your mouth is **menthol**. Menthol works the same way capsaicin does—by interacting with a portal found in your nerve cells. You have both hot and cold sensors. Capsaicin stimulates the hot sensor and menthol stimulates the cold one. You also have these sensors in your skin, which is why some skin creams include menthol. A chemical that looks almost identical to menthol is menthone. We're going to ask you to spot the difference between menthol and menthoe, but we'll make it easier by showing you a more simplified way to draw these chemicals.

Here are the molecules with every atom shown. Here they are with some of the C's and H's taken out.

If you look at the simplified drawings, it is easy to see the difference. The simplified drawings are missing seven C's, plus the H's attached to the C's. Sometimes, chemists will even omit (leave out) the CH_3's. They will just leave a "stick" hanging off, and trust that you know there must be a carbon atom at the end, plus its three hydrogens. The simplified drawings are a little hard to read at first, but you can see why they are used.

The phytochemicals produced by the peppermint plant are intended to be insecticides that deter bugs from biting the leaves. These chemicals are found in cells that form tiny hair-like structures on the leaves and stems. These cells produce over a dozen phytochemicals as well as oils. When these hair-like structures touch something, they burst open, releasing their chemicals.

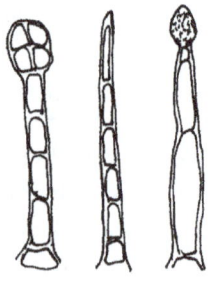

Any type of sweetener, be it sugar or a sugar substitute, does the same thing in your mouth. Not only do you have hot and cold sensors, you also have sweet sensors. Any molecule that is able to stimulate the sweet sensors in the nerve cells on your tongue will cause those nerve cells to send an electrical signal to your brain. Your brain will interpret this signal as "tastes sweet."

The most popular all-natural sweetener is **stevia**. It comes from a plant called stevia. The name of the plant comes from the name of a Spanish botanist, Petrus Stevus. The molecule that triggers your sweet sensors is very large and complicated, so we're not going to bother you with it. Basically, it has a glucose molecule bonded to a larger molecule that you can't digest. So the sugar triggers the sensor but then passes through your intestines as fiber.

The native peoples of Brazil and Paraguay have been using stevia as a sweetener for at least 1,500 years.

Another all-natural, no-calorie sweetener that is rising in popularity is **monk fruit**. Monk fruit is estimated to be well over 100 times as sweet as sucrose, so just a pinch will sweeten your drink. As with stevia, the monk fruit has molecules that have sugars attached to them, but the sugars stay on the molecule and are not able to be digested by our bodies. So they tingle our sweet sensors, then pass through our systems undigested. Monk fruits have been grown in southern China for centuries and were known as Luo Han Guo. The name "monk fruit" comes from the fact that the monks of Luo Han used this plant for herbal medicine.

A natural, but processed, sweetener is **xylitol**. "Xylo" is from the Greek word for "wood." The "-ol" on the end tells you that part of the molecule is classified as an "alcohol," meaning that is has an "-OH" bonded to each carbon atom. Xylitol occurs naturally in some fruits and vegetables but in very small quantities. Large scale production of xylitol was first proposed during World War II when sugar was in short supply. Xylitol can be made from almost any type of vegetation, including a lot of the waste (corn husks, rice hulls, bark, etc.) left over from other processes. The key molecule is xylan, which is found between the cells of almost all plants. The xylan undergoes several chemical processes to turn it into xylitol. Though not often used in kitchens, xylitol is widely used in sugar-free gum and candy.

If you want to learn about some artificial sweeteners, you can read the following page. Oddly enough, all four of these sweeteners were discovered by accident!

"Oops! I discovered a sweetener..."

Saccharin *(SACK-a-rin)* was discovered by accident in 1879 by a chemist named Fahlberg. He spilled an unknown chemical on his hand and didn't bother to wash up before dinner (not a recommended lab procedure). He noticed the bread he was eating was unusually sweet and by tasting residues on his clothes and hands (also not a recommended lab procedure), figured out that the sweetness had come from the chemical. He later named this chemical saccharin. By 1907, saccharin began to be used as a sugar substitute. In the 1960s it began to be used by the soft drink industry for diet sodas. Two sweeteners that contain saccharin are "Sweet 'N Low®" and "Sugar Twin®."

Despite the controversy over whether saccharin is safe for human consumption, not enough evidence has piled up to have it taken off the "generally recongnized as safe" list. This does not mean it is the right choice for everyone. Some people seem to have negative responses after eating saccharin.

Aspartame was discovered by accident in 1965 by a chemist named Schlatter. He was working on a project aimed at discovering new treatments for stomach ulcers. One of the steps in the process was to make something called "aspartyl-phenylalanine methyl ester." He accidentally spilled some on his hand and later licked his finger as he reached for a piece of paper (not a recommended lab procedure). He noticed a sweet taste and decided to test the chemical to see if it would sweeten his coffee. It did, and the rest, as they say, is history.

Cyclamate was discovered at the University of Illinois in 1937. A chemistry graduate student named Michael Sveda was working on developing a new fever-reducing medicine. Back then, it was considered normal to smoke cigarettes while working, so Michael was smoking as he was working with his chemicals. He laid his cigarette on the lab bench in order to use both hands to move some equipment. When he put the cigarette back into his mouth, he noticed a sweet taste and realized he had gotten cyclamate on it. The lab went on to patent cyclamate as an artificial sweetener and it was very popular in the U.S. until the FDA banned it in 1970. A rat study from 1969 had shown that 8 out of 240 rats came down with bladder cancer when forced to consume the amount of cyclamate you'd get if you drank 550 cans of diet soda every day. Some countries still use cyclamate.

Sucralose was discovered by accident in England. A British sugar company was doing research on sucrose. They were adding chlorine atoms to the sucrose molecules. They sent samples of this chlorinated sugar to one of their employees for testing. This particular employee was not a native English speaker and he misunderstood what they wanted. He thought they wanted it "tasted" instead of "tested." Well, he tasted it... and found out that this substance is about a hundred times sweeter than sugar.

NOTE: None of this information is intended to endorse any product.

NOTE: Don't forget to check the YouTube playlist one last time!

* *

Comprehension self-check

1) What is the primary role of the yolk and the white in a developing egg?

2) What is the role of the blastoderm spot on the yolk?

3) What is the name of the emulsifier found in eggs?   a) albumen   b) lecithin   c) collagen   d) carrageenan

4) Name two things that will cause egg proteins to unfold:

5) In meringue, the egg white proteins gather around _____ _____.

6) When a protein unfolds and straightens, we say it has been: d_____.

7) What plant is most often used to make vegetable shortening?

8) Which is curved, a saturated fatty acid or an unsaturated fatty acid?

9) Which contains double bonds, saturated fatty acids or unsaturated fatty acids?

10) Which will be a solid at room temperature, saturated fatty acids or unsaturated fatty acids?

11) When vegetable oils are hydrogenated, do they become more solid or less solid?

12) "Trans" means "across." In trans fats, _____ atoms are across from each other.

13) Will you find trans fats in many products today?

14) Is a gelatin aspic sweet?   What do people put into aspics?

15) Name three animal sources of collagen used for gelatin:

16) Cream is made of three-legged fat molecules called:

17) In whipped cream, the fat molecules collect around the outside of _____ _____.

18) Name two thickeners made from seaweed:

19) What thickener is found in fruit?

20) TRUE or FALSE? Jell-O® was an instant success.

21) Name two plants that are used to make root starch thickeners:

22) Does theobromine taste sweet?

23) How much of a cacao bean is fat?   a) 5%   b) 15%   c) 50%   d) 95%

24) Which of these is NOT designed to be toxic to pests?   a) caffeine   b) theobromine   c) pectin   d) menthol

25) Cacao is considered a superfood because it contains:

26) Sucrose can be separated into glucose and fructose by _____ or _____.

27) Sugar syrup that has had its sucrose turned into glucose and fructose is called _____ syrup. (pg. 111)

28) When broken sugar and protein molecules form nonsense molecules this is called the _____ reaction.

29) Which of these is mixed with sugar to make fondant?   a) gelatin   b) starch   c) fat   d) vegetable oil

30) Which of these was not discovered by accident?   a) aspartame   b) stevia   c) saccharin   d) sucralose

ACTIVITY 6.1 And now for the bad news...

Sugar is so sweet and so much fun to eat, but sadly, the truth is that it's really not very good for us. Scientific studies have shown that eating more than a certain amount of added sugars per day can significantly affect your health, even if you are young. So how much is too much? Many health agencies recommend no more than about 36 grams (9 teaspoons) for an adult male and no more than 25 grams (6 teaspoons) for an adult female.

Food products that come in boxes and bags are required to tell you how many grams of sugar have been added. Some products have a surprising amount of sugar. For example, a can of soda ("pop") has about 40 grams (10 teaspoons) of sugar. Can you imagine eating 10 spoons of sugar? Yet when you drink a can of sweetened soda, that's how much sugar you are taking in.

Find out how many grams of added sugar you eat in a day. Keep track of at least three days because, as we all know, some days you eat more or less of various foods. (The days don't have to be all in a row.) Under each day, list the name of the food and how many grams of sugar. (Don't count natural sugars, like apples.)

DAY 1	DAY 2	DAY 3
TOTAL:	TOTAL:	TOTAL:

ACTIVITY 6.2 How many carbon atoms?

At the very end of this chapter we learned how chemists draw carbon-based molecules. They assume everyone knows that at every vertex (corner or intersection), or at the end of every empty stick, there is a carbon atom and enough hydrogen atoms to give the carbon the four bonds it wants. Figure out how many carbons are in each of these molecules. (Your answer will include any carbons that are shown as "C".)

ACTIVITY 6.3   REVIEW CROSSWORD PUZZLE

These words are from previous chapters.

ACROSS
1) (OH-) is called the _____ ion.
3) A polyphenol found in tea.
5) The structure in a plant cell where starch molecules are stored.
6) The molecule found in peppers that triggers the hot sensors in our tongue.
12) A "hydrogen ion" is the same thing as a _____.
14) A glycerol molecule with three fatty acids attached to it.
15) Liquids that have particles evenly dispersed through them. (Milk is an example.)

ACROSS (continued)
17) If you cut this molecule in half you get two retinol molecules.
22) A microscopic ball of protein found in milk.
23) The "amine" group contains two hydrogen atoms and a _____ atom.
24) The molecule that makes leaves green.
25) Six carbon atoms and six hydrogen atoms will form a _____ ring.
26) Most of a corn seed is made of _____, which stores starch plus some other nutrients.
27) This element (type of atom) allows molecular bridges between long chains, such as glutenin and gliadin.
30) A substance that can hold on to both fat molecules and to water molecules.
31) Carbon dioxide and _____ are waste products created by yeast during the bread making process.
33) A potato is classified as this plant part.
34) The enzyme that can cut apart a molecule made of glucose and fructose.

DOWN
1) When milk fat globules are put through a screen that makes them all the same size they have been _____.
2) This type of nutrient is essential to life and was discovered by studying deficiency diseases.
4) Another name for vitamin C is _____ acid.
7) These are the things that enzymes put together or tear apart. (Bet you've forgotten what they are called!)
8) The enzyme that can take apart the sugar molecule found in milk.
9) The enzyme that can disassemble the long chains of glucose molecules that form starch.
10) These are usually blue or green and are put into certain cheeses to create distinctive flavors.
11) This pigment molecule produces both red and blue colors, depending on the surrounding pH.
13) When bacteria turn pyruvate molecules into lactic acid this process is called _____.
16) The milk protein used for paint and glue.
18) A tiny particle with a negative charge.
19) A long chain of glucose molecules that we can digest.
20) Most of a bean seed is made of the "seed leaves" which are also called the _____.
21) A long chain of glucose molecule that we CAN'T digest.
27) In a salt water solution, the salt is the _____.
28) A one-celled fungus that is used to make bread rise.
29) What is known as the universal solvent?
32) Casein is the main protein in milk. All the other smaller proteins are known collectively as _____.

ACTIVITY 6.4   Questions, questions, questions...
Imagine you are shopping with a 5-year old, and as you walk through the aisles they do what all kids do—ask a zillion questions. Here are a few of them. Can you answer them? (They aren't one word answers!)

1) Why is goat butter white?
2) Is corn a vegetable?
3) Why don't oil and water mix?
4) Why do people say carrots are good for your eyes?
5) How to beverage companies get so much fizz into their cans of carbonated drinks?
6) How can cows live on nothing but grass?
7) Why is brown rice better for you than white rice?
8) What's so bad about gluten?
9) Why does crispy brown bread crust smell and taste so good?
10) What's wrong with trans fats?
11) Is margarine better than butter?
12) Can any vegetable be purple?

ACTIVITY 6.5   Final installment of "Chew It Over," a group game to be played during a meal

Here is another round of questions for you to use at a mealtime that you share with family or friends. These questions relate to the topics we learned about in this chapter. Again, you can use these questions in a variety of ways. You can be the quiz master and determine who gets which questions, or you can cut the questions out of the book and put them into a bag or bowl and let people choose a question randomly. The answers on are the back of this page.

CHAPTER 6: DESSERT

1) Guess the weight of the world's largest chocolate bar (as recorded by Guinness):

a) 500 lbs (225 kg)   b) 2,300 lbs (1,043 kg)
c) 12,770 lbs (5790 kg)   d) 47,500 lbs (21,500 kg)

CHAPTER 6: DESSERT

2) We get our word for chocolate from the Aztec word "xocolatl." Guess what it means.

a) sweet food   b) dark medicine
c) bitter water   d) edible gold

CHAPTER 6: DESSERT

3) Guess the melting point of chocolate:

a) 80° F (26.6 °C)   b) 93° F (33.9° C)
c) 100° F (37.7° C)   d) 120° F (48.8° C)

CHAPTER 6: DESSERT

4) Try to guess each person's favorite kind of pie. (For those who don't like pie, you can guess another favorite food.)

CHAPTER 6: DESSERT

5) Can you guess which country consumes more chocolate per person that any other country?

a) United States   b) France
c) Italy   d) Switzerland

CHAPTER 6: DESSERT

6) WHICH STATEMENT IS FALSE?
1) Grape juice is much lower in sugar than soda.
2) Cats cannot taste sugar.
3) Sugar can be used as a preservative.
4) Most sugar comes from beets, not sugar cane.

CHAPTER 6: DESSERT

7) WHICH STATEMENT IS FALSE?

1) Benjamin Franklin sold chocolate in his print shop.
2) George Washington told his wife never to serve chocolate.
3) Soldiers during the American Revolutionary War were sometimes paid with chocolate.

CHAPTER 6: DESSERT

8) WHICH STATEMENT IS FALSE?
1) An egg can have two yolks.
2) The color of a hen's earlobes predict what color eggs they will lay.
3) Brown eggs are more nutritious than white eggs.

CHAPTER 6: DESSERT

9) How many ways can you think of to cook an egg?

CHAPTER 6: DESSERT

10) WHICH STATEMENT IS FALSE?
1) Older hens will lay eggs with thinner shells.
2) "Free range" hens always live outdoors.
3) Fresh eggs will sink in a glass of water.
4) Some chickens lay blue eggs.

CHAPTER 6: DESSERT

11) INTERESTING FACT:
Did you know that the less sugar you eat, the more sensitive you will be to its taste? Once your brain adjusts, a lower amount of sugar will taste just as sweet.

CHAPTER 6: DESSERT

12) INTERESTING FACT:
White chocolate doesn't contain any theobromine. The only part of the cocoa bean that is used is the fat (the "cocoa butter").

1) c   2) c   3) b
5) d   6) #1   Grape juice is very high in sugar.
7) #2
8) #3   Brown and white eggs are identical on the inside, with no nutritional differences.

INDEX

acetic acid	55
actin	89
agar	108
albumen	99
alkaline	22
alpha carbon	25
alpha-galactosidase	86
amino acid	40, 51
amylopectin	48
amylase	49
amyloplast	83
amylose	48
anthocyanin	65, 77
antioxidants	67
ascorbic acid (vit C)	69
aspartame	114
aspic	105
atom	3
atropine	96
bacteria	43, 74
baking powder	54
baking soda	54
base	22
bean pod	82
benzene ring	18
benzoic acid	19
beriberi	69
beta-carotene	39, 65
betalain	77
bicarbonate ion	54
blastoderm	99
butter	29, 30
cacao	109
caffeine	17, 110
calcium carbonate	99
calcium phosphate	24
capsaicin	81, 82, 96
caramelization	110, 111
carbon dioxide	15
carbonation	17
carcinogen	19
carotenoids	66
carrageenan	108
casein	25
celiac	52
cellulase	75
cellulose	74
chalaza	99
cheese	42
chlorine	6
chlorophyll	65, 66
chocolate	109
chromoplast	88
chymosin	41
coagulation	101
cobalamin	70, 71
collagen	106
colloid	28
compound	3
cotyledon	83
curcumin	67
curdling milk	41
cyclamate	114
cysteine	26, 50
denaturing	100
dissolving	7, 16
disaccharide	11
disulfide bond	50
density	6
DNA	64, 92
egg	99
egg white	100
electron	4
electron microscope	1
element	2
emulsifier	61, 62
emulsion	61
endive	73
endosperm	84
enzyme	9, 10, 75
ethanol	45
fatty acids	24, 61
fermentation	44-46
fiber	74
flavonoids	67
fondant	111
fructose	11
Funk, Casimir	70
galactose	26
gliadin	50-52
glucose	11, 47
gluten	50, 105
glutenin	50
glycerol	24
glycemic index	49
glycine	26
glycoprotein	41
glyphosate	52
green bean	82
hemagglutinin	85
heme	65, 90, 92
hemoglobin	90
hydrogenation of oils	103, 104
homogenization	23
hydrogen bonding	5, 16
hydrogen ion	21
hydrophobic	61
hydroxide ion	21
iceberg	73
Jell-O	105
Keeney, Calvin	82

lactase	27	raffinose	85
lactic acid	38	rennet	41
Lactobacillus	23	resveratrol	67
lactose	27	retinol	36
leaf	63-65	riboflavin (B2)	70
leavening	54	rickets	71
lecithin	99	RDA	73
lectins	84	RNA	64
leghemoglobin	92	saccharin	114
lettuce	63, 73	salad dressing	61, 62
Lind, James	68	salmon	90
Lunin, Nickolai	69	salt crystals	6, 8
lycopene	66	saturated fats	91, 102
Maillard reaction	53, 110	scurvy	68
mayonnaise	62	sodium	6
menthol	112	sodium acetate	23, 55
meringue	100	soybean	92
microbiome	74	solanine	87, 96
mint	112	solute	7, 15
mixture	16	solvent	7, 15
monosaccharide	11	spinach	73
moonflower	96	starch	48, 83, 84, 101, 108
morning glory	87	stevia	113
myofibril	89	stomata	63
myoglobin	89	substrates	10
muscle	89	sucrase	8, 9
myosin	89	sucralose	114
niacin	70	sucrose	9, 111
nightshades	96	sugar	110, 111
nitrogen	17, 83	sulfur	83
neutron	4	sweet potato	87, 88
neutralize	22	tannins	67
oil bodies	83	tartaric acid	55
omega-3	91	theobromine	109, 110
ovary	82	thiamine (B1)	69
pasteurization	23	thylakoid	65
pectin	108	trans fat	104
peptidase	26	triglyceride	24, 38, 103
peptide bond	25, 26	tuber	89
Periodic table	2	turmeric	67
pH	22	Tyndall effect	28
phenol	67	unsaturated fats	91
phospholipid	65	vanillin	18
phosphoric acid	20	vitamin A	36
phosphorus	20	vitamin B1	69
phytochemicals	66	vitamin B2	70
phytohemagglutinin	84	vitamin B3	70
phytonutrients	66	vitamin B6	70
plastid	88	vitamin B12	70, 71
plumules	83	vitamin C	68, 69
polar molecules	5, 16	vitamin D	71, 72
polypeptide	26	vitamin E	72
polyphenols	667	vitamin K	72
potassium benzoate	19	whey	28, 42
potato	87, 88	xanthophyll	65
preservative	19	xylitol	113
protein bodies	83	yam	87
proton	4	yeast	45, 46, 64, 93
pyridoxine (B6)	70	yellow #5	18
pyruvate	43, 44	zonulin	52

ANSWER KEY

CHAPTER 1

Review questions:
1) Protons, neutrons, electrons   2) Protons and neutrons   3) Sharing of electrons.   4) 2
5) Water is polar because one side of the molecule is slightly positive and the other side is slightly negative.
6) Hydrogen bonding
7) The atoms in salt (sodium and chlorine) have an electrical attraction to water molecules.
8) the salt   9) substrates
10) Enzymes either tear things apart or they put things together. Each enzyme does only one job.
11) sucrose   12) sucrase   13) glucose and fructose   14) water
15) Water is necessary for tearing apart molecules. The OH and the H are used to patch up ragged ends.

CHAPTER 2

Review questions:
1) The water molecules pull at the solute molecules, destroying any crystals, and trap the solute molecules in water "cages." (Answers can vary, as long as they express this general idea.)
2) no   3) cold   4) oxygen   5) up   6) nose
7) Benzoic acid is harmful to microorganisms but not to people. Benzoic acid might be harmful to people if it turns into benzene, a substance that might be carcinogenic. (Answers can vary as long as they contain this idea.)
8) microorganisms in their food
9) People like a like some sour "tang" as well as the sweet taste.
10) proton   11) hydroxide ion   12) acid   13) basic   14) alkaline   15) 7
16) water, a salt   17) Bacteria (germs) are killed.
18) Large fat globules are broken up so that all the fat globules are made the same size.
19) casein   20) peptidases   21) 20   22) NH_2 23) glucose and galactose
24) sucrose   25) solution and colloid, milk is colloid

2.1
1) solutes   2) hydrogen bonding   3) pesticides   4) 6   5) 5   6) benzene   7) raw   8) carcinogens
9) potassium   10) vitamin C   11) phosphoric acid   12) hydroxide ion   13) water, salt   14) alkaline
15) pasteurized   16) homogenized   17) glycine   18) glycerol   19) micelles   20) peptidases
21) function   22) FDA   23) simple sugars   24) fatty acids   25) water   26) gas, liquid
27) pressure   28) Canada   29) bananas   30) glasses

Answers to "Interesting Fact about Ice Cream and Root Beer"
1) mashed potatoes   2) August 6   3) twelve   4) sorbet   5) 5 gallons
6) akutaq, moose fat, seal oil, snow, berries, fish   7) the churn   8) St. Louis   9) America, Australia, Norway
10) sassafras   11) medicine   12) sarsaparilla

2.2
1) seal   2) black rhino   3) sheep   4) rabbit   5) hippo   6) goat   7) whale   8) donkey   9) horse   10) wallaby

CHAPTER 3

Review questions:
1) carbon   2) smashes them, breaks them up   3) water   4) bacteria
5) curdles it   6) lactose   7) beta-carotene
8) any fruit or vegetable that is orange or yellow   9) A, retina   10) casein
11) micelles, negative, glycoproteins   12) curdled, vinegar, lemon juice, rennet, cows, bacteria
13) whey   14) cottage cheese   15) bacteria, mold   16) glycolysis, pyruvates
17) lactic acid, fermentation   18) carbon dioxide, yeast   19) ethanol
20) fungus   21) starch, glucose   22) amylase, yes   23) glutenin, gliadin
24) cysteine   25) zonulin   26) Maillard   27) an acid

3.1

Avenin	Buttermilk	Carboxyl	Disulfide
Enzyme	Fermentation	Gliadin	Hydrogen
Index	Juice	Kitchen	Lactate
Mitochondria	Negative	Orange	Phosphate
Question mark	Retinal	Starch	Threonine
Unload	Vacuole	Whey	oxygen
Yarg	Zein		

3.2

<u>Stinky cheeses:</u>
1) Limburger   2) Camembert   3) Epoisses (Aye-pwahz)
4) Taleggio   5) Stinking Bishop   6) Roquefort
7) Raclette   8) Mimolette   9) Casu Marzu   10) Parmesan

<u>Key words:</u>
PROTEIN, BUTTERMILK, LIQUID, ZONULIN, REFLECTS, ETHANOL, RANCID, GOATS

CHAPTER 4

Review questions:
1) electrons   2) emulsifier   3) b   4) 2, 1, 3   5) seeds, c   6) stomata, guard
7) palisade   8) FALSE   9) phospholipids   10) chlorophyll   11) beta-carotene
12) heme   13) magnesium   14) carotenoids   15) anthocyanin   16) curcumin
17) antioxidants   18) tannic acid   19) To protect them from pests and from sun damage.
20) scurvy, beriberi   21) d   22) a   23) c   24) A and D   25) D
26) K   27) coenzyme or cofactor   28) cellulose, cellulase   29) no
30) To push food along, and also to feed the "good" bacteria that live in your gut.
31) spinach   32) iceberg   33) spinach   34) endive   35) spinach
36) spinach   37) spinach   38) iceberg   39) no   40) endive

4.2

<u>Crossword:</u>
ACROSS: 1) tannins   3) Funk   5) riboflavin   6) turmeric   11) hydrophobic   13) rickets
         14) spongy   15) photosynthesis   17) coenzyme   18) fruit   23) emulsifier
         24) pyridoxine   26) resveratrol   27) cellulose   28) antioxidant   29) vacuole
DOWN: 2) starch   3) fiber   4) thiamine   7) microbiome   8) chloroplast   9) polyphenols
         10) phospholipid   12) ascorbic   14) stomata   16) carotenoids   19) cuticle
         20) heme   21) cellulase   22) magnesium   25) nucleic

CHAPTER 5

Review questions:
1) peppers (or hot peppers, or chili peppers), (a) heat   2) animals-yes, birds-no   3) ovary
4) To provide nutrients for the new baby plant.   5) a) proteins   6) triglycerides
7) glucose   8) corn, rice, wheat, oats   9) clump together   10) carries oxygen molecules
11) lectins, b) sugar strings   12) cook at high heat   13) raffinose   14) -ose, -ase
15) no   16) potatoes   17) TRUE   18) potatoes   19) sweet potato
20) FALSE   21) actin and myosin   22) myoglobin ("heme" also acceptable)   23) myofibril
24) carbon monoxide   25) from its food source   26) c) hydrogen atoms   27) bacteria, nitrogen
28) yeast   29) Maillard   30) legume

CHAPTER 5 (con't)

5.1
1) j, phytohemagglutinin   2) k, solanine   3) m, starch granules   4) d, capsaicin
5) c, cotyledon   6) i, plumule   7) e, endosperm   8) l, alpha-galactosidase
9) n, root tuber   10) g, green   11) a, actin   12) o, oxygen
13) a, astaxanthin   14) f, fatty acids   15) h, nitrogen

5.2
1) amino acid   2) glucose   3) enzyme   4) micelle   5) phospholipid   6) cellulose   7) chlorophyll
8) beta-carotene   9) phenol   10) emulsifier   11) oligosaccharide   12) myoglobin   13) triglycerides
14) starch

CHAPTER 6

Review questions:
1) nutrition for the developing bird embryo   2) The blastoderm grows into the chick embryo.
3) b) lecithin   4) heat and physical force (beating, whipping)   5) air bubbles   6) denatured
7) soybean   8) unsaturated fatty acid   9) unsaturated fatty acids   10) saturated fatty acids
11) more solid   12) hydrogen   13) no   14) no, meat, eggs, vegetables
15) cow hooves, pig skins, fish   16) triglycerides   17) air bubbles   18) agar and carrageenan
19) pectin   20) False   21) arrowroot, cassava   22) no   23) c) 50%
24) c) pectin   25) vitamins, minerals, phytochemicals such as flavonoids and polyphenols
26) heating, enzymes   27) invert   28) Maillard   29) a) gelatin   30) b) stevia

6.2
1) 4   2) 10   3) 12   4) 7   5) 11   6) 6   7) 22   8) 18   9) 18   10) 27

6.3
ACROSS: 1) hydroxide   3) tannin   5) amyloplast   6) capsaicin   12) proton   14) triglyceride
15) colloids   17) betacarotene   22) micelle   23) nitrogen   24) chlorophyll   25) benzene
26) endosperm   27) sulfur   30) emulsifier   31) ethanol   33) tuber   34) sucrase
DOWN: 1) homogenized   2) vitamin   4) ascorbic   7) substrates   8) lactase   9) amylase
10) mold   11) anthocyanin   13) fermentation   16) casein   18) electron   19) starch
20) cotyledons   21) cellulose   27) solute   28) yeast   29) water   32) whey

6.4
1) Goat butter is white because of the way goats store beta-carotene. Their cells chop the beta-carotene in half to make two retinols, and they store the retinols. Retinol does not reflect as much yellow light as beat-carotene.
2) No, not really. Only vegetative (nothing related to reproduction) parts count, such as leaves, stems and roots. Corn is a seed, so an ear of corn is technically a fruit. Anything with a seed is a fruit.
3) Water is a polar molecule and oil is non-polar. Oil molecules are "hydrophobic" and want to stay away from water. Only polar molecules that have a positive and negative side will be attracted to water molecule.
4) Carrots have beta-carotene which our bodies can chop in half to make molecules of retinol. Retinol is made into retinal which fits into a protein in the light sensing area of our eye.
5) They use very high pressure and low temperatures (just above freezing).
6) Cows have a special stomach called the rumen, which is full of bacteria. The bacteria digest the plants that the cow eats. The cow then absorbs the nutrients as well as digesting and absorbing the bacteria themselves.
7) Brown ride still has the hull around the seed. The hull is full of phytonutrients and vitamins, and it has fiber which is good for the intestines. Brown rice takes longer to digest which gives it a lower glycemic index.
8) Gluten isn't "bad." The problem is that some people's bodies see the string of proteins as a piece of information instead of simply nonsense amino acids. The information triggers the intestine cells to become "leaky."
9) Crispy browning is cause by Maillard reactions, when sugars, fats and proteins get broken apart and then reform into bizarre nonsense molecule that happen to smell and taste good.
10) Trans fats have hydrogen atoms on opposite side of the long molecule. For some reasons, when they are in this shape they tend to get stuck in places they should not, such as along the insides of our blood vessels.
11) People used to think that saturated fats like butter were worse for you than unsaturated products like margarine, but now in the 21st century, new studies have shown that saturated fats are not a problem. So go ahead and eat butter!
12) Basically yes, just about any vegetable can be purple. We rarely see purple beans or corn or peppers, but they do exist.

BIBLIOGRAPHY

Books

How Baking Works by Paula Figoni; published by John Wiley & Sons, Inc., 2004. ISBN 0-471-26856-9
How Did We Find Out About Vitamins? by Isaac Asimov; published by
Food Science; The Biochemistry of Food and Nutrition by Kay Yockey Mehas and Sharon Lesley Rodgers; published by Glencoe McGraw Hill, 1997. ISBN 0-02-647647-9

Websites   (those I remembered to write down)

Water, salt and sugar
https://www.education.com/activity/article/Make_Homemade_Glue/
https://en.wikipedia.org/wiki/Carbonated_water
https://en.wikipedia.org/wiki/Sodium_hydroxide
https://www.stevespanglerscience.com/lab/experiments/instant-hot-ice/
https://www.scientificamerican.com/article/lift-ice-cubes-with-chemistry/
http://dryiceinfo.com/

Milk
http://drcate.com/raw-milk-why-mess-with-udder-perfection/
http://factslist.net/2013/01/why-hippos-milk-is-pink-10-random-facts-about-hippos/
http://ansci.illinois.edu/static/ansc438/Milkcompsynth/milkcomp_table.html
https://en.wikipedia.org/wiki/Sheep_milk
https://en.wikipedia.org/wiki/Donkey_milk
https://www.smithsonianmag.com/science-nature/seven-most-extreme-milks-animal-kingdom-180956588/
https://www.livestrong.com/article/512165-enzymes-to-digest-milk/
http://scitoys.com/ingredients/casein.html (explanation of casein micelles)
http://www.food-info.net/uk/protein/milk.htm
https://www.researchgate.net/figure/7240581_fig1_Figure-1-Major-secretory-pathways-in-mammary-epithelial-cells-during-lactation-A
https://dairyextension.foodscience.cornell.edu/sites/dairyextension.foodscience.cornell.edu/files/shared/Composition%20of%20Milk.pdf
http://ansci.illinois.edu/static/ansc438/Milkcompsynth/milksynth_proteinmajor.html
http://vlab.amrita.edu/?sub=3&brch=63&sim=158&cnt=1

Butter and cheese
http://oklahoma4h.okstate.edu/aitc/lessons/dairy/butter.pdf
https://www.cooksillustrated.com/how_tos/5690-butter-additives
https://www.youtube.com/watch?v=lfeXuK8pbFw
https://well.blogs.nytimes.com/2016/10/14/since-milk-is-white-why-is-butter-yellow/
https://en.wikipedia.org/wiki/Beta-Carotene
www.kickassfacts.com/top-10-stinkiest-cheeses-in-the-world/
http://www.seriouseats.com/2014/03/a-brief-guide-the-best-stinky-cheeses.html
https://www.thedailybeast.com/why-we-love-stinky-cheese
https://www.thespruce.com/describing-cheese-like-a-pro-591154
http://intro.bio.umb.edu/OLLM/112s99/dairyBacteria.html
https://microbiologylearning.weebly.com/common-bacterial-scents.html
https://www.khanacademy.org/science/biology/cellular-respiration-and-fermentation/variations-on-cellular-respiration/a/fermentation-and-anaerobic-respiration
http://www.differencebetween.com/difference-between-lactate-and-vs-lactic-acid/
https://en.wikipedia.org/wiki/Lactic_acid

Wheat and gluten
http://sciencegeist.net/of-bread-and-old-age/
https://www.ncbi.nlm.nih.gov/pubmed/8783747
https://en.wikipedia.org/wiki/Zonulin
https://www.ncbi.nlm.nih.gov/pmc/articles/PMC3384703/

Amino acids
https://www.thoughtco.com/amino-acid-373556
https://en.wikipedia.org/wiki/Amino_acid

Capsaicin
https://en.wikipedia.org/wiki/TRPV1
https://en.wikipedia.org/wiki/Capsaicin

Meat
https://www.exploratorium.edu/cooking/meat/INT-what-meat-color.html
https://www.todayifoundout.com/index.php/2010/04/the-red-juice-in-raw-red-meat-is-not-blood/
https://en.wikipedia.org/wiki/Myoglobin
https://faq.impossiblefoods.com/hc/en-us/articles/360034767354-How-do-you-make-heme-

Beans
http://www.saveseeds.org/biography/keeney/index.html
https://www.saturdayeveningpost.com/2009/05/green-bean-backgrounder/
https://en.wikipedia.org/wiki/Kidney_bean
https://en.wikipedia.org/wiki/Phytohaemagglutinin
https://www.frontiersin.org/articles/10.3389/fpls.2020.00809/full
https://www.sciencedirect.com/topics/agricultural-and-biological-sciences/protein-bodies
https://www.hsph.harvard.edu/nutritionsource/anti-nutrients/lectins/
https://en.wikipedia.org/wiki/Lectin
https://en.wikipedia.org/wiki/Raffinose
https://en.wikipedia.org/wiki/Oligosaccharide
https://en.wikipedia.org/wiki/Leghemoglobin
Leghemoglobin: https://www.youtube.com/watch?v=jG9tO1_20aA
https://altonbrown.com/recipes/aquafaba-meringues/

Starch storage in cotyledons
https://www.nature.com/articles/s41598-019-40424-w

Starch and iodine
http://www.webexhibits.org/causesofcolor/6AC.html
https://en.wikipedia.org/wiki/Iodine%E2%80%93starch_test

Sweet potatoes/yams
https://ncsweetpotatoes.com/sweet-potatoes-101/difference-between-yam-and-sweet-potato/
https://en.wikipedia.org/wiki/Sweet_potato
https://www.npr.org/sections/thesalt/2013/01/22/169980441/how-the-sweet-potato-crossed-the-pacific-before-columbus

Plastids (conversion between chloroplasts, amyloplasts)
https://www.frontiersin.org/articles/10.3389/fpls.2021.692024/full

Whipped cream
https://www.seriouseats.com/the-science-of-whipped-cream-butter-creme-fraiche
https://bcachemistry.wordpress.com/tag/whipped-cream/
https://en.wikipedia.org/wiki/Whipped_cream
https://foodretro.com/the-science-of-whipped-cream-and-butter/
https://sci-toys.com/ingredients/whipped_cream.html
https://www.thekitchn.com/whats-the-difference-between-half-and-half-light-cream-whipping-cream-and-heavy-cream-73203

Egg experiment blue dye goes through pores
https://www.scientificamerican.com/article/bring-science-home-chick-breathe-inside-shell/

Eggs
https://extension.psu.edu/programs/4-h/get-involved/teachers/embryology/teacher-resources/supporting-subject-matter/the-egg/the-parts-of-the-egg
https://www.exploratorium.edu/cooking/eggs/eggscience.html
https://en.wikipedia.org/wiki/Egg_white
https://curiodyssey.org/wp-content/uploads/2019/03/Science-Experiment-Egg-Dissection.pdf

Lemon
https://www.ncbi.nlm.nih.gov/pmc/articles/PMC7020168/

Pie crust
https://qz.com/931472/how-to-make-the-perfect-pie-crust-according-to-science/

Gelatin
https://en.wikipedia.org/wiki/Jell-O
https://en.wikipedia.org/wiki/Gelatin

Red dye
https://www.amenclinics.com/blog/brain-health-guide-red-dye-40/

Caramelization
https://www.scienceofcooking.com/caramelization.htm
https://sciencenotes.org/carmelization-chemistry-why-sugar-turns-brown/

Cacao
https://en.wikipedia.org/wiki/Theobroma_cacaor
https://en.wikipedia.org/wiki/Theobroma_cacao
https://www.youtube.com/watch?v=g-F8zL3Ufi4
https://www.youtube.com/watch?v=kVdor-5kM0o
https://www.thespruceeats.com/what-is-cocoa-butter-3376444
https://www.eatthis.com/chocolate-facts/

Corn syrup
https://en.wikipedia.org/wiki/Corn_syrup
https://en.wikipedia.org/wiki/Inverted_sugar_syrup

Peppermint
https://www.acmicrob.com/microbiology/peppermint-and-its-functionality-a-review.php?aid=19955
https://sciencemeetsfood.org/cool-flavors-mint/

Sweeteners
https://en.wikipedia.org/wiki/Stevia
https://foodinsight.org/everything-you-need-to-know-about-monk-fruit-sweeteners/
https://www.lakanto.com/pages/monk-fruit

Enzymatic browning
https://extension.purdue.edu/4h/documents/volunteer%20resources/science%20made%20easy/brown%20apples.pdf
https://www.mombrite.com/apple-oxidation-science-experiment/

How polyphenols help your cells
https://www.ncbi.nlm.nih.gov/pmc/articles/PMC2634513/

SUPPLEMENTAL ACTIVITIES

Most of the supplies you need for these activities are household items that you either already have on your shelf or can be easily purchased at your local grocery store, pharmacy, department store, or craft store.

The only specialty items you might want to order ahead of time are:

1) Lugol's solution (iodine solution)
   I order from www.homesciencetools.com, but you can also get it on Amazon.
   A standard 2% solution should be fine.

2) Eye droppers (pipettes), either reusable or disposable
   You can get plastic or glass—your choice. Available at both places I listed above.

3) Agar ("agar agar") **powder**

4) Optional: Sodium alginate, calcium lactate (or calcium chloride) for making popping boba in chapter 6. Read about it first, then decide whether you want to invest in these materials. The cost in USD is around $20-30.

5) Optional: Self-adhesive jewel stickers for craft activity 3G. View the activity and decide if you want to use them. If so, search online at Amazon or a craft store website. (However, I found mine at my local Dollar Store.)

* *

If you have only a paperback copy of this book and would like a digital copy of the project pattern pages that you can use for printing (either home computer or at a copy and print service), can find one at this web address:

https://ellenjmchenry.com/printable-pages-for-DYD

(This page also has a link to a file for the activity pages in the student text.)

IF YOU WOULD LIKE SUGGESTIONS FOR MORE INVOLVED RESEARCH PROJECTS (THE KIND THAT CAN BE USED FOR SCIENCE FAIRS) CHECK OUT THIS RESOURCE:

https://www.juliantrubin.com/fairprojects/food/foodchemistry.html
https://www.juliantrubin.com/fairprojects/food/food_preservation.html

The suggestions are grouped by age levels: elementary, middle and high school.

CHAPTER 1

ACTIVITY IDEA 1A   EXPERIMENT: HOW MANY DROPS OF WATER WILL FIT ON A PENNY?

You will need:
- an eye dropper for each student
- a penny for each student
- cups of water (students can share)

What to tell the students:
    In this chapter you read about hydrogen bonding between water molecules. This attraction between water molecules is what allows water to form droplets. In this activity you will observe the shape of a single droplet, then you will see how many drops of water will fit on top of a penny before gravity finally overcomes the strength of the hydrogen bonds.

What to do:
1) Give each student a penny, an eye dropper and access to a cup of water.
2) Tell them to put one single droplet onto the penny. Look at it from the side. How close is it to being a sphere?
3) Then give them the challenge of slowing putting additional droplets onto the penny, carefully counting as they go. Make sure the eye dropper (or the droplet being added) doesn't touch the penny or the droplet on the penny. When the penny is full of water look at it from the side. How tall is the drop?
4) At some point (over 30 drops perhaps) the giant droplet will finally give way and let go, gushing out over the penny and onto the table.
5) The students will likely want to repeat this procedure several times to see if they can add more drops.

ACTIVITY IDEA 1B   GAME: WATER DROP RACES

You will need:
- an eye dropper for each student
- a plastic straw for each student
- waxed paper
- pencil and paper for each student
- round object for tracing a large circle on the paper (small plate, for example)
- tape
- cups of water (students can share)
- for game #2: a tablet or stiff piece of cardboard the size of a sheet of paper

What to tell the students:
    In this activity you will play games with water droplets.

What to do for GAME #1:   RACE TO THE CENTER
1) Give each student a piece of paper and a pencil. Pass around the round object and have the students place it in the center of their paper and trace around the edge. (If you have a large group of students, you can either prepare the circle pages yourself ahead of time, or use an object that you have multiples of, such as a set of saucers.)
2) Tell the students to lay the waxed paper on top of their sheet of paper and use a few pieces of tape to secure the waxed paper to the table so it won't slide around.
3) Have the students fill their droppers and place 12 drops of water around the edges of their circle (outside the line). Imagine the circle to be a clock, and put a drop on each "hour."
4) Now they are ready to play the game. The goal is to get all of their water drops together in the center of the circle, forming one large drop. Use the plastic straws to move the water drops around on the waxed paper. Ready, set, go! First person to get all their droplets together in the center wins the game.

What to do for GAME #2:   RACE DOWN THE ROAD
1) Give each student a tablet or piece of cardboard and have them tape a piece of paper to it. Then have them draw a curved "road" down the page. Tape the waxed paper on top of the paper.
2) Put a drop of water at the top of the curved road and then tip and tilt the road, trying to keep the water droplet inside the track while it rolls downward.

ACTIVITY IDEA 1C   EXPERIMENT: NO STIRRING ALLOWED!

You will need:
- a cup of water for each pair of students
- food coloring

What to tell the students:
All molecules, including water molecules, are in constant motion. The molecules themselves vibrate internally, and they also wander about, bumping into other molecules. Although we can't see water molecules, we can witness their motion by putting a drop of dye into the water. The dye is also made of molecules that are moving. We can watch as the dye and water molecules move around and get mixed together. This will take some patience, as it is not a fast process.

What to do:
1) Give each pair of students a cup of water to observe.
2) Tell the students that they are NOT to disturb the cup in any way. No touching the cup! Warn the students that they will need to be patient and watch the cup for several minutes.
3) Put a single drop of food coloring into each cup.
4) The drop of food coloring will stay in place for a surprisingly long time. However, it will begin to swirl a bit and flow into other areas of the cup. The students' patience will be the limiting factor in this experiment. You might want to get this experiment started, then take a break and come back to it 5 or 10 minutes later.
5) Eventually, the dye will completely disperse into the water, making the water a homogeneous pale color.

ACTIVITY IDEA 1D   EXPERIMENT: OBSERVING CRYSTALS

You will need:
- salt and sugar
- small pieces of black paper
- Optional: Epsom salt crystals, a specialty salt such
- as sea salt or Himalayan salt, or a special type of sugar
- Optional: include some sand, to represent another type
- of tiny crystal that is easily identified
- magnifiers (ideally 20x or higher, but minimum 10x)

What to tell the students:
In this activity, you will see that different types of salt and sugar can be identified by their crystal shapes.

NOTE: If you do not have magnifiers, you can use the following page.

What to do:
1) Give each student a magnifier, a piece of black paper, and access to the crystals you are supplying.
2) Have them take a pinch of each substance and place it on their black paper. Tell them to observe carefully.
3) Extension: After they observe all the crystals, have them clean off their black paper, then give them a pinch of a mystery crystal (don't let them see which container you used) and see if they can correctly identify what it is.

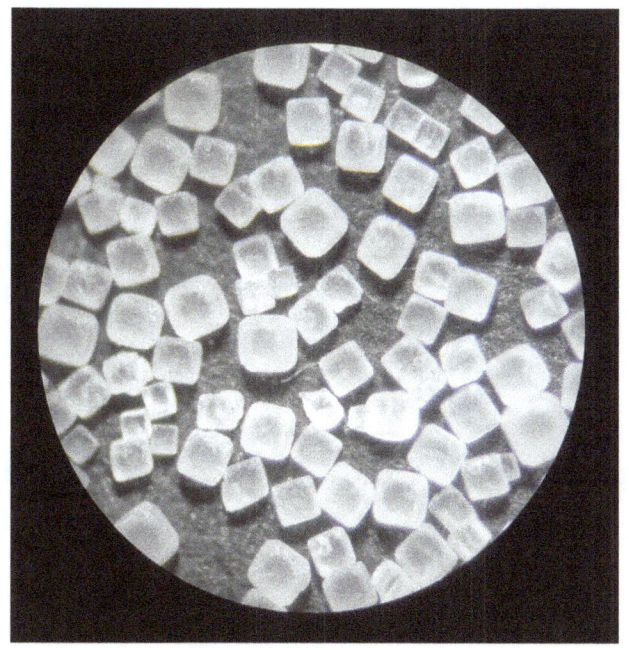
table salt   40x

Epsom salt 20x

Himalayan salt   20x

raw cane sugar 40x

sand   40x

ACTIVITY IDEA 1E EXPERIMENT: USE SALT TO LIFT AN ICE CUBE

You will need:
- salt
- ice cubes
- several short pieces of string
- bowl or glass of water
- Optional: a marker to darken the end of one string

What to tell the students:
   In this experiment you will use salt to lift an ice cube with a piece of string. Salt lowers the freezing point of ice by interfering with ice's crystal structure. Salt will cause ice to melt at a temperature less than 0˚ C (32˚ F).

What to do:
1) Give each student, or pair of students, a bowl or glass of water with a few ice cubes floating in it and two pieces of string. (Optional: Use the marker to darken the end of one string so you can keep track of which is which.) Have the students lay a piece of string across two ice cubes.
2) Tell them to shake a little salt on top of one cube but not the other.
3) Let the cubes sit for a few minutes.
4) Lift the strings. If the cubes have been sitting long enough, the string that was salted should be stuck to the ice cube and will be able to lift it. The unsalted string won't be stuck to the cube.
5) Explain that the salt caused enough melting at the surface of the cube to allow the string to sink into the cube just a bit. The cold temperature of the rest of the cube then causes this melted area to quickly refreeze, trapping the string in the ice.

ACTIVITY IDEA 1F EXPERIMENT: GROW LARGE SALT AND/OR SUGAR CRYSTALS

You will need:
- white granulated sugar
- non-iodized salt
- water (distilled if possible, but use tap water if distilled not available)
- pan, stove, stirring spoons
- glass jars
- pieces of string, or wooden sticks such as Popsicle sticks or bamboo skewers
- optional: food coloring

Time required: At least a week to grow the crystals

What to tell the students:
   In this activity, you will watch the growth of crystals of salt and sugar. We will be using a type of solution called a "super saturated" solution. This means that the solvent (in this case, water) is holding the maximum amount of solute that it possibly can. Our solutions will have a lot of dissolved salt or sugar and some of it will come out of solution and stick to a string or stick you will provide, and will hopefully grow to become a very large crystal.

What to do:
1) Make your super-saturated solutions. The same technique is used for both salt and sugar. You heat water in a pan until it is almost boiling. You don't need a rolling boil, just extremely hot water. You keep adding salt (or sugar) (not both!) until no more will dissolve into the water. Eventually, there will be so much dissolved salt or sugar that what you are adding simply falls to the bottom of the pan.
2) Turn off the heat and let it solution cool. (Add some food coloring if you'd like, but this is not necessary.)
3) Pour only the clear top part of the solution (no solids from the bottom) into your jar or glass.
4) Dip the end of your string or stick into the solution. Let this sit until dry.
5) The bit of dried solute on your string or stick will act as the "seed crystal" that will be the starting point for the larger crystals that will form.
6) Rig a way to dangle the string or stick in the solution so the tip is well into the solution but not touching the bottom or sides.
7) Let this sit for several days, or even several weeks. Watch as the crystal on the end gets larger and larger.

EXTENSION FOR ACTIVITY 1F: While you have this solution available, try this bonus experiment. Put a few drops of each solution onto a smooth surface. Jar lids will work well, or small plates. Allow some drops to air dry. Use a hair dryer or fan to force other drops to dry quickly. Then compare the crystals. Expected result is that the crystals that dried slowly will be larger and have more impressive geometric shapes. Forming nice crystals takes time.

ACTIVITY IDEA 1G   EXPERIMENT: PICKING UP SALT WITH A BALLOON

You will need:
- salt (any type)
- small plate
- one balloon per student or pair of students
- Optional: pepper or sugar

What to tell the students:
In this activity you will see that salt crystals carry an electrical charge. Things that carry an electrical charge are attracted to other objects that are electrically charged.

What to do:
1) Put a spoon of salt into the dish or plate. Blow up the balloon.
2) Before rubbing the balloon against anything, hold the balloon over the salt and see what happens. (Expected result is that nothing will happen. If the balloon happens to have an electrical charge already, some salt might jump up onto it.)
3) Rub the balloon against your hair.
4) Hold the balloon over the salt and see what happens. Expected result is that many grains of salt will jump up and stick to the balloon. You are likely to hear a sprinkling sound as they hit the balloon. NOTE: If you are doing this experiment in a room that has a lot of humidity (during hot summer day, for example) you might get less spectacular results. Best results happen with low humidity.
5) Extensions: Have students brush salt off and see if the balloon still has enough static to draw more salt. Try the same thing with pepper or sugar.

ACTIVITY IDEA 1H   EXPERIMENT: BREAKING APART WATER MOLECULES

You will need:
- clear glass of water
- 9V battery
- tiny pinch of salt
- two short wires, if possible (and something to strip the plastic off the ends)

What to tell the students:
In this activity you will split water molecules into oxygen gas and hydrogen gas. The gas bubbles will be very small, but they will be visible. Will we see more oxygen bubbles or more hydrogen bubbles, or the same number of each?
Think about the chemical formula: H_2O. What does the "2" mean?

What to do:
1) Strip both ends of the wires. If you don't have an actual wire stripper, you can use a pair of scissors or a knife to put a cut in the rubber coating. Once cut, the bit of rubber on the end should slide right off leaving a bit of bare wire.
2) Attach one wire to each terminal on the battery.
3) Put a tiny pinch of salt into the water.
4) Put the wires into the water. If you don't have wires, just turn the 9V battery upside down and submerge just a tiny portion of the end with the terminals. (Warning: The battery could get hot if you keep the terminals wet for several minutes.)
5) Watch for bubbles to form. The wire, or terminal, with more bubbles will be hydrogen because the formula, H_2O, tells us that every time you split a molecule, you get two hydrogens and only one oxygen.

ACTIVITY IDEA 1I   MOLECULE MAT for chapter 1

You can do this activity with a wide variety of materials. You can use edibles such as soft candies or bits of fruit or veggies, or you can use inedibles such as colored play dough or clay. If you happen to have molecule building kits (the sturdy plastic type) you might be able to match enough parts to the molecules listed here, although you might run into problems with the balls not having the correct number of holes or not having enough of some colors.

TIP: If you want to be able to reuse the pattern pages, you can slide them into plastic sleeves ("sheet protectors").

What you will need if you are using toothpicks and small, soft objects:
- toothpicks that are cut in half (or you can try short pieces of uncooked spaghetti) (Play dough shown in photo.)
- a copy of the following pattern page for each student
- small, soft objects to represent each type of element (20 H's, 11 O's, 13 c's, 4 Na's, 4 Cl's)

What to tell the students:
In this activity you will be making some of the molecules you read about in the chapter. The toothpicks will represent the electrical attractions that keep the atoms together. If you are good at making models and would like an extra challenge, make sure that the four sticks coming out of your carbon atoms are as far apart from each other as possible.

What to do:
1) Put your chosen materials inside the boxes on the left side of the page (or in small dishes if they won't fit inside the boxes). Toothpicks can be set in a dish, or simply in a pile, within the students' reach.
2) Let the students work on their own as much as possible.
3) For students who are keeping a portfolio of their work, take a photo of their paper with all the finished molecules on it.

For students who finish early (if you are working with a group), try one of these ideas:
1) Have them use the diagram in their book to make fructose.
2) Have them disassemble their glucose molecule and then see if they can reassemble it from memory. (Make sure they can't see the diagram on the mat!)

EXTENSION: If you are working with a group, you might want to have the students pick up their finished glucose molecules and place them all side by side in line. Then ask them to create bonds between the molecules. (Remember, an OH comes off one molecule and an H off the other to create an H_2O.)

NOTE ABOUT DOUBLE BONDS IN CARBON RINGS:
You will notice that in many carbon rings, the bonds in the ring alternate between one line and two lines. Carbon always wants to make four bonds, so normally we picture it as having four lines coming out of it. With this single/double line structure, the carbons have only three lines coming out of them, suggesting they are only making three bonds. In reality, what is going on is that these bonds are going back and forth between single and double so fast that is impossible to tell which is which. Therefore, the carbons think they have four bonds enough of the time that they are content and the molecule will be stable. However, because the possibility of adding an atom exists, this opens the possibly of an enzyme being able to add a new part to the molecule by changing those carbon intersections into a standard 4-bond situation.

Remember, if you have this book in paperback form, you can download a digital copy of the Molecule Mat by going to http://ellenjmchenry.com/printable-pages-for-DYD. The mat pattern is included in the printable pages download file.

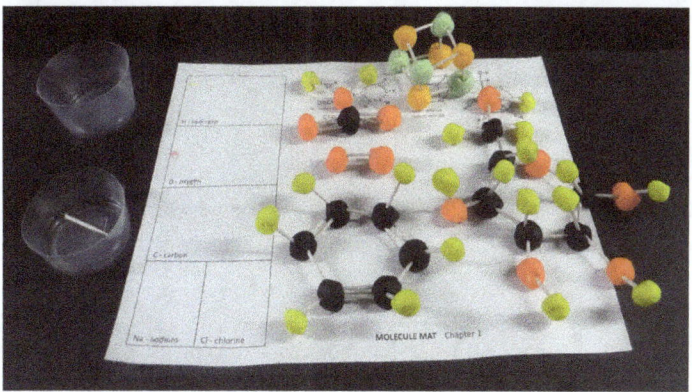

MOLECULE MAT Chapter 1

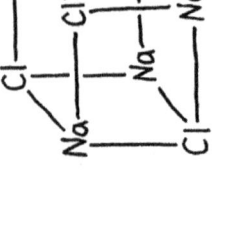

C₆H₁₂O₆ glucose

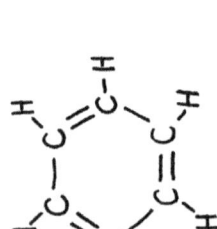

NaCl table salt
(sodium chloride)

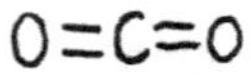

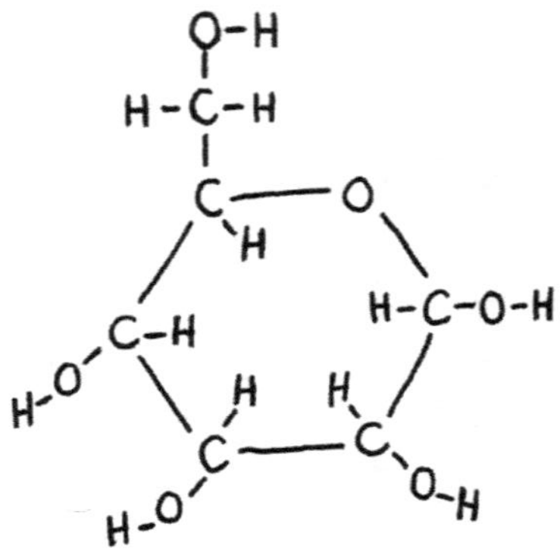

C₆H₆ benzene ring

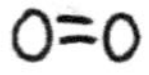

H–O–H   H₂O water

O=C=O   CO₂ carbon dioxide

O=O   O₂ oxygen

H - hydrogen	
O - oxygen	
C - carbon	
Na - sodium	Cl - chlorine

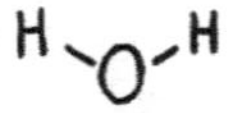

CHAPTER 2

(NOTE: Even though this chapter introduced pH, we are going to wait to do pH experiments until chapter 4 when we read about anthocyanin, which can be used as a pH indicator.)

ACTIVITY IDEA 2A   EXPERIMENT: "UNSHAKE" A CAN OF SODA (CARBONATED BEVERAGE)

You will need:
- one or more cans of carbonated beverage (TIP: Cans of seltzer water are often cheaper than flavored sodas.)

What to tell the students:
   You know what happens when you shake a can of soda pop and then open it. In this activity you will try to reverse the shaking process so the beverage does not explode out of the can when you open it.

What to do:
1) Shake a beverage can.
2) Tap the top of the can with 5-10 strong, firm taps. If you can't get a firm enough tap using your finger, you can use an object like a spoon.
3) Open the can slowly.

   When you shake a can of soda, you are causing many tiny bubbles to form all along the inside of the can. These tiny bubbles will be the gathering point for larger bubbles to form if the pressure inside the can is released. By tapping the can, you are forcing these tiny air bubbles hanging to the inside of the can to float to the surface. Once these tiny air bubbles are gone from the inside of the can, the can has been restored to its original state.

EXTENSION: Try it with both warm and cold cans. (Cold beverages hold dissolved carbon dioxide better than warm ones, and should therefore make less mess.)

ACTIVITY IDEA 2B   SNACK: MAKE A CARBONATED DRINK WITH "DRY ICE"

You will need:
- dry ice
- cups
- fruit juice of your choice

What to tell the students:
   Dry ice is frozen carbon dioxide gas. If we put a piece of dry ice into a glass of juice, the carbon dioxide should diffuse into the juice, making it fizzy. Dry ice is much colder than regular ice. Be careful not to touch it!

What to do:
1) An adult should break up the dry ice into cube-sized pieces. (Be careful not to touch the dry ice as it is much colder than regular ice and could cause injury to skin.)
2) Let each student hold their cup and watch the dry ice fizz and create lots of vapor. Their drink will look like a magic potion while the dry ice is melting. Tell them not to drink their juice until the piece of dry ice has completely melted.

ACTIVITY IDEA 2C   EXPERIMENT: RISING AND FALLING RAISINS (a.k.a. "DANCING" RAISINS)

You will need:
- raisins
- tall, clear glass (so you can see the action going on inside)
- seltzer water

What to tell the students:
   In this experiment you will observe the lifting power of carbon dioxide bubbles. Raisins are more dense that water and will naturally sink. When put into carbonated water, the wrinkled skins of the raisins provide lots of places where microscopic "nucleation sites" can form. These sites will attract CO_2 and form bubbles. The bubbles sticking to the raisin will give it enough lift to be able to rise to the surface. At the surface, the bubbles pop and the raisin sinks. Then the cycle repeats.

What to do:
1) Fill the glass with seltzer water and add about two dozens raisins.
2) Observe the raisins becoming covered with bubbles and rising to the surface. Watch as the bubbles pop and thus disappear at the surface, allowing the now bubble-free raisin to sink back to the bottom.

EXTENSION: Try some other objects, such as popcorn kernels, candy sprinkles, dry lentils. Do they form as many bubbles? Why or why not? (Do smooth objects have fewer nucleation sites?) Also, you could try comparing a clear plastic glass to a clear glass made of glass. Which has more bubbles forming on the sides? Might the texture of the plastic (even though it appears to be smooth) be less smooth than glass?

ACTIVITY IDEA 2D   FOR AMBITIOUS EXPERIMENTERS: TRY THE "HOT ICE" EXPERIMENT

If you like challenging experiments, try making "hot ice" using baking soda and vinegar. Instead of reading how to do it, it's much better to watch. Check out the video on the playlist, or use a video search engine and the key words "hot ice experiment." If your hot ice doesn't work, at least you have a batch of sodium acetate crystals to look at. Just allow the solution to evaporate completely (and slowly) and you'll have some interesting crystals.
NOTE: Sodium acetate is used in many reusable hand warmers. If someone in your class has them, you could prepare them (they "reset" when boiled) and let the class witness how fast they go from flexible to hard. "Hot ice" forms inside the packet.

ACTIVITY IDEA 2E   COOKING ACTIVITY: MAKE "MILK GLUE" (CASEIN GLUE) or MILK "PLASTIC"

In this activity, you will "curdle" milk. We will explain curdling in the next chapter. You can wait to do this activity until after chapter 3, or you can use it now if you need more activities for this chapter.

You will need:
- milk (any type—fat content won't alter results)
- white vinegar
- baking soda
- water
- pan, stirring spoon, strainer (NOTE: Some sources recommend straining using a funnel lined with a coffee filter.)

What to do:
1) Warm 1.5 cups (350 ml) milk in a pan. Once it is warm, add 3 teaspoons (5 ml) vinegar. (Measurements don't have to be exact.)
NOTE: If you are working in a classroom, you can do this process without heating and it should still work reasonably well.
2) Keep heating (if you are heating) and stirring until you see the curds and whey separate.
3) Strain out the curds. (You won't need the watery whey.) The curds will form a lump.
4) Put the lump of curds back into the pan and add a little water and a teaspoon of baking soda.
5) Heat this mixture until it bubbles. Turn off heat and let it cool.
6) Once cool, this mixture should make a stiff paste that can be spread using a brush or craft stick.

EXTENSION: The lump of casein (from step 3) can also be used as "milk plastic." Simply mold the casein into whatever shape you'd like, then let it dry. The dried "plastic" can be painted with acrylic paints.

NOTE: Making casein paint is more difficult than making glue. If you want to try paint, search online for instructions. It is mainly use by people in the antique furniture business.

ACTIVITY IDEA 2F   "JUST FOR FUN" ACTIVITY: FOOD COLORING IN MILK

This activity is often done with younger children, but older kids enjoy it, too. Older students will be able to appreciate the chemistry of the materials.

You will need:
- milk (higher fat milk works best)
- food coloring
- plates
- dish soap
- cotton swabs

What to tell the students:

In this activity you will watch interactions between polar and non-polar molecules. Milk is mostly water, which is polar. It has microscopic non-polar fat globules suspended in it. The food coloring you will add to the milk will let you see molecular motion in the milk that would otherwise be impossible to see. Soap molecules have both a polar end and a non-polar end, so they can interact with both water and fat. When you touch a drop of soap to the center of the plate, the soap will begin to disperse in the water because of its polar ends. Then its non-polar ends will begin to be attracted to fat molecules. Further more, the negatively charged end of the soap molecules are attracted to the milk's protein molecules. So there are molecules coming and going all over the place as they chase after molecules that they'd like to be associated with. The food coloring simply goes along for the ride and allows us to see all the motion going on at the molecular level.

What to do:
1) Pour enough milk into the plate to completely cover the bottom.
2) Put a few drops of food coloring into the milk. DO NOT STIR!
3) Touch the tip of a cotton swab to the soap.
4) Touch the soapy end of the swab to the center of the milk.

NOTE: The effect in the milk will "run out" after a few touches of the swab. If they want to see the effect again, you'll need to rinse the plate and put in fresh milk.

EXTENSION: Some online sources suggest laying a piece of paper on top of the swirly designs to "capture" the design. Pull if off carefully and lay it flat to dry.

ACTIVITY IDEA 2G   EXPERIMENT: TESTING FOR THE TYNDALL EFFECT (COLLOIDS)

You will need:
- various liquids in clear glasses (For example: water, whole milk skim milk, apple juice, water with food coloring, water with various powders mixed in, such as flour, baking powder, cornstarch, etc.)
- a laser pointer
- a flashlight

What to tell the students:

In this experiment you will test liquids for the Tyndall effect that you read about at the end of this chapter. The Tyndall effect occurs in colloids. Solutions will let light pass right through because the dissolved particles are so small. Colloids have larger particles which will block the light.

What to do:
1) Start with a clear glass of water. Shine the flashlight and the laser pointer through and observed what happens to the light.
2) Mix a little milk into the water and shine the light again. Any change? Add a little more and try it again.
3) Shine the light through a glass of pure milk.
4) Try the light through various other liquids. How many are colloids? Try mixing some powders (powdered fruit juice, jello powder, powdered milk, or whatever you have on hand) into water and see which end up being solutions and which end up being colloids.

ACTIVITY IDEA 2H   GAME: "THE GLYCINE GAME"   (pattern pages are part of the digital download at http://www.ellenjmchenry.com/printable-pages-for-DYD)

You will need:
- copies of the following pattern pages (see printing directions at the bottom of each page)
  If card stock is not available, use regular paper.
- scissors
- Optional: paper and pencil for each player (for writing down answers to questions)

How to set up:
1) Choose one of the card sets to use in the game. If you want to do review of facts learned in the chapter, print the REVIEW cards. If your students have mastered the basic concepts of the chapter, you can choose BONUS INFO cards. (Or print both of them and play the game twice!)
2) Cut apart the molecule cards and the quiz cards. Trim the playing board.
3) Shuffle the molecule cards and place them face down on the square in the center of the board. Shuffle your chosen set of quiz cards and place them face down on the rectangle in the center of the board.

How to play:
1) The goal of the game is to be the first player to use the molecule cards to build a glycine molecule (as shown on board).

2) Players take turns drawing cards. If the card is a quiz question, the reader of the card will be the one player who is not allowed to benefit from a right answer since, obviously, they are seeing the answer. The other players write down their guess as to the correct answer. When the correct answer is read aloud, each player who had written the correct answer gets to take a molecule card. If the card is simply a question about what you ate (from the BONUS INFO card set), then the reader of the card may also give an answer, if applicable, and receive a card.
NOTE: For the BONUS INFO cards, players are not expected to already know the answers. They are supposed to think about any information they might know that would help them to eliminate wrong answers, then take a guess. If they guess wrong, that's okay. They learn when the hear the answer. (If that question comes up again, hopefully they will remember the answer.)

3) The role of the H_2O cards: Water is the "universal solvent," meaning that because it is a strongly polar molecule, it exerts a pull on many types of chemical bonds. When you draw an H_2O card, it will dissolve one of your chemical bonds. This means that you must choose one of your cards and return it to the bottom of the draw pile. You also return the H_2O card to the bottom of the draw pile.

4) The first player to make a complete glycine molecule wins the game.

ACTIVITY IDEA 2I   EXPERIMENT: EFFECTIVENESS OF VARIOUS PRESERVATIVES IN BREAD

You will need:
- several slices of bread from different sources that use various preservatives, or use no preservatives
(Calcium propionate and sodium proportionate are the most common preservatives. Some natural breads use a chemical derived from cherries. In my class, the cherry preservative worked amazingly well.)
- sandwich-sized plastic bags
- marker to write on the bags

What to tell the students:
   In this chapter we met a few preservatives. In this experiment you will find out the names of various preservatives used in commercial bread and you find out which seem to be most effective at preventing mold.

What to do:
1) Place one slice of bread from each loaf into a plastic bag. The bags will keep the bread from drying out and will speed up the molding process.
2) Read the labels on the bread bags and find the names of the preservatives. (When in doubt, type the chemical name into an Internet search engine to find out what it does.) Label the bags so you remember which loaf each slice came from. You might also want to write which preservatives were listed on the bread bag.
3) Let the bags sit for at least a week, until obvious mold spots appear. Observe the bags each day and keep a written record of what happens in each bag.
4) Discuss results. Which bread was the most resistant to mold? Think of other factors that might have contributed to the rate of molding. (ex: water content, types of grains used, salt or sugar content)

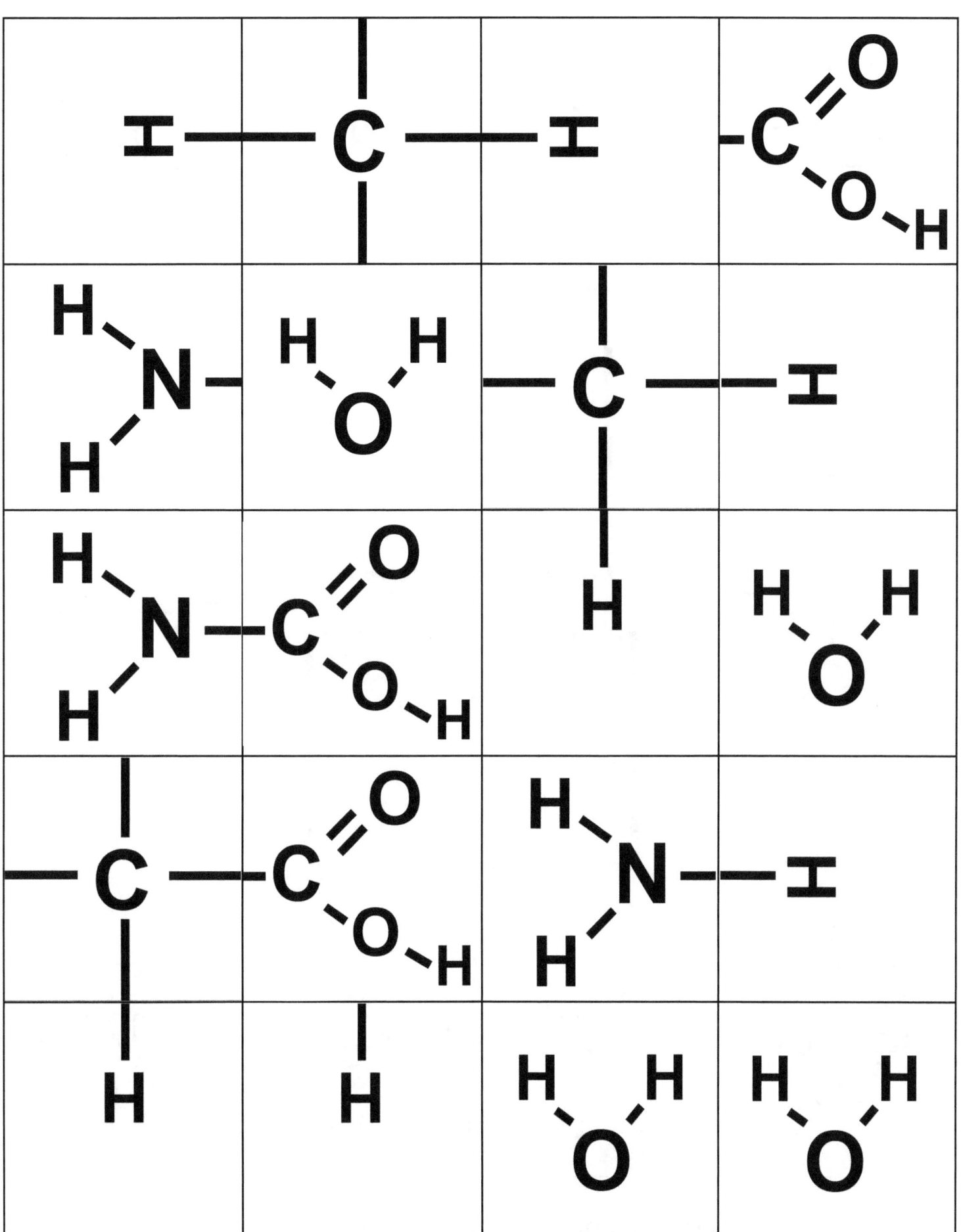

PRINT ONTO CARD STOCK / PRINT ONE PAGE FOR 2 PLAYERS, PRINT 2 PAGES FOR 3-4 PLAYERS

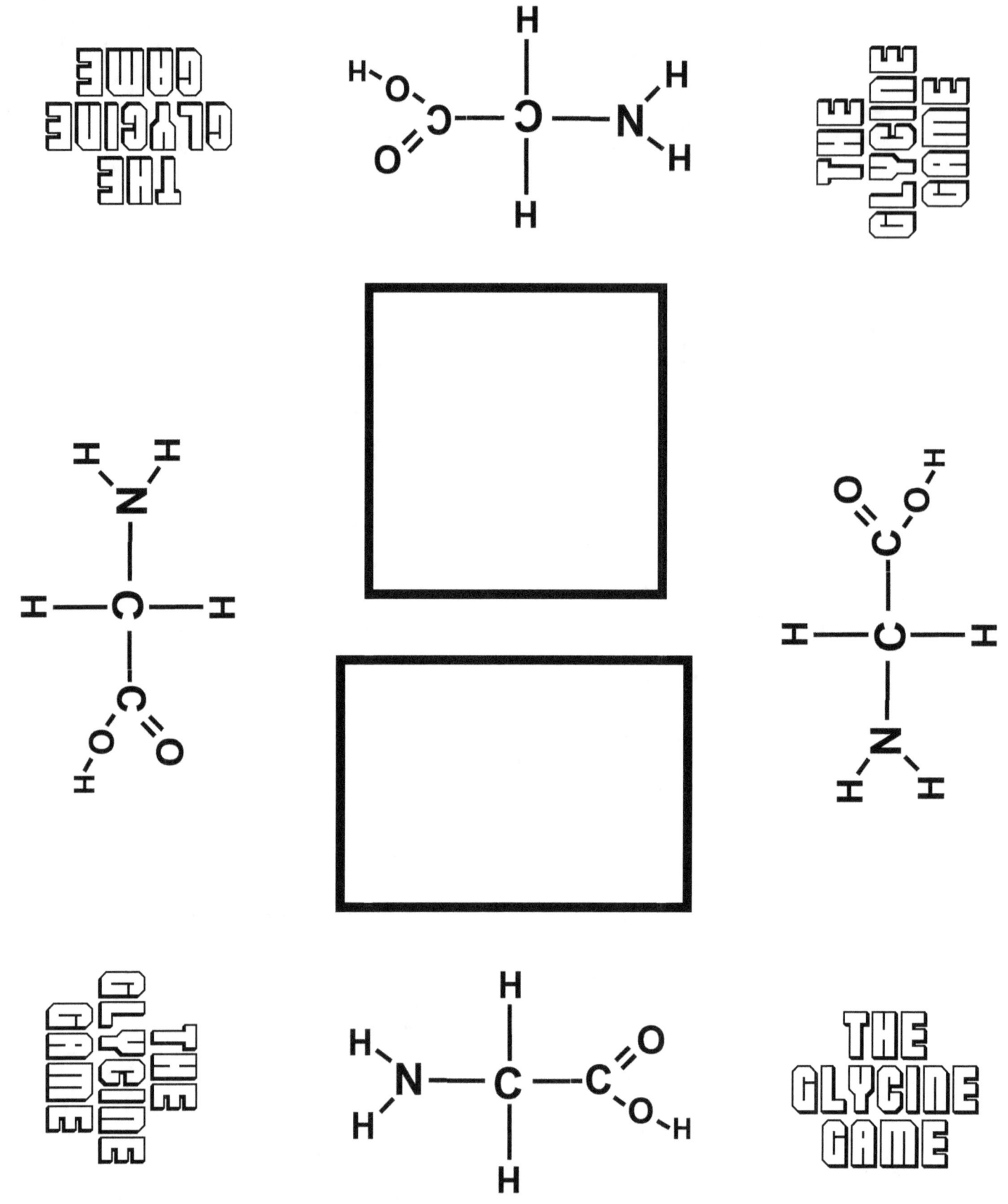

What number is neutral on the pH scale? Answer: 7	What is the special name for the attraction between water molecules? Answer: hydrogen bonding	TRUE or FALSE? What we think of as "taste" comes largely from our sense of smell. Answer: True
TRUE or FALSE? The amount of benzene in a can of soda is dangerous. Answer: False	What is OH- called? a) hydrogen ion b) hydronium ion c) hydroxide ion Answer: C	TRUE or FALSE? A hydrogen ion is the same thing as a proton. Answer: True
TRUE or FALSE? If you let a nail sit in a glass of Coke, the phosphoric acid in the Coke will dissolve the nail. Answer: False	Salt water is a: a) solution c) colloid b) mixture d) compound Answer: A	Milk is a: a) solution c) colloid b) mixture d) compound Answer: C
When a glucose molecule is attached to a fructose molecule, what do you get? Answer: sucrose	What is the name of the enzyme that breaks apart the sucrose molecule? Answer: sucrase	What is the special name for the carbon atom at the center of an amino acid molecule? Answer: alpha carbon
How many fatty acid chains are on a triglyceride molecule? (Hint: Think about the word.) Answer: 3 ("tri" means 3)	What is the name of the enzyme that breaks apart the lactose sugar molecule? Answer: lactase	Peptidase enzymes break apart what type of molecule? a) fats b) proteins c) sugars Answer: B
TRUE or FALSE? Trigylceride fat molecules in milk float around freely Answer: FALSE, they are found inside spherical membranes.	Famous scientist Louis Pasteur discovered this simple sugar. Answer: Galactose	What is casein? a) Glue that is made from milk. b) A protein found in milk. c) A type of paint made from milk. Answer: B

REVIEW CARDS   PRINT ONTO CARD STOCK / ONE COPY PER GAME (up to 4 players)

Take a card if you have consumed any food or drink today that contains casein. Remember, casein is the protein found in milk. Casein foods include milk, cheese, yogurt, and ice cream. Butter and cream have had most of their casein removed, so they are not really casein foods.	Food allergies are usually not a good thing, but today they will earn you a card. Name your food allergy or allergies and then take a card.	Can you guess which of these plant parts has the highest amount of protein? a) roots b) leaves *c) seeds d) flowers
Everyone (except the card reader) loses a card. Choose one of your cards and return it to the bottom of the draw pile.	If we could gather all the sodium and chlorine atoms in your body, and turn them into salt, can you guess how many shakers they would fill? a) half a shaker b) one shaker *c) three shakers d) ten shakers	Nuts are full of amino acids! Take a card if you have eaten any type of tree nut today or yesterday. Tree nuts include walnuts, almonds, pecans, pistachios, brazil nuts, pine nuts, macadamias. hazelnuts, and cashews. NOTE: Peanuts are NOT tree nuts.
Salt can be extracted from sea water. Can you name the country that has the saltiest sea in the world? a) Turkey b) Lebanon c) Egypt *d) Israel	Time to trade! Each player must choose one of their cards and hand it to the player on their left.	The ancient Romans were famous for the stinky, salty sauce that they used as a condiment (like ketchup). The sauce was left to ferment (rot) for months. Can you guess the main ingredient of their stinky sauce? *a) fish b) beans c) cabbage d) garlic
The healthiest (mineral-rich) drinking water in the world is generally considered to be water that comes from what source? a) lakes *b) glaciers c) the ocean d) underground wells	Beans are a great source of protein. If you have eaten any beans today or yesterday, take a card. Ex: green beans, kidney beans, black beans, cannellini bean, Fava beans, Lima beans, refried beans	Can you guess who drinks more milk than anyone else in the world? The citizens of: *a) Norway b) Netherlands c) Germany d) France
Dairy cows produce over 80% of the world's milk. The runner up, with 14% is: a) sheep b) goats *c) buffalo d) camels	Can you guess where the world's largest dairy farm is located? They milk over 90,000 cows every day! *a) Arabia b) Korea c) Russia d) South Africa	Can you guess the most popular dairy cow in the United States? a) the brown and white Guernsey *b) the black and white Holstein c) the tan and white Jersey
Which type of milk will look slightly blue if you shine a light through it? a) raw milk b) whole milk c) 2% fat milk *d) fat-free skim milk	Which of these plant sources is NOT used to make a milk substitute? a) almonds b) oats *c) peanuts c) coconuts	This question is about something you read in the chapter. Which one of these substances do you think has the lowest pH number? a) water b) apple juice c) milk *d) lemon juice

BONUS INFO CARDS   PRINT ONTO CARD STOCK / ONE COPY PER GAME (up to 4 players)

ACTIVITY IDEA 2.J   "MOLECULE MAT" for chapter 2

You will need:
- a copy of the following pattern page for each student
- toothpicks
- the materials you used for the Molecule Mat in chapter 1 (or try something different if you want to)

What to do:
1) Put your chosen materials inside the boxes on the left side of the page (or in small dishes if they won't fit inside the boxes). Toothpicks can be set in a dish, or simply in a pile, within the student's reach.
2) Let the students work on their own as much as possible.
3) The molecules are likely to be large enough that all four will not fit onto the page. You can tell your students to build two then recycle the items, or give them an extra blank sheet and have them build several molecules on it.
4) For students who are keeping a portfolio of their work, take a photo of their paper with all the finished molecules on it.

TIP: For students who finish early (if you are working with a group) have them bond their glycine to their alanine. (Remove the OH from the COOH, and one of the H's from the NH_2. The OH and H will form a water molecule.) If they need another challenge, have them build caffeine, shown here on the right.
 (We will build this after the last chapter, but doing it twice is fine.)

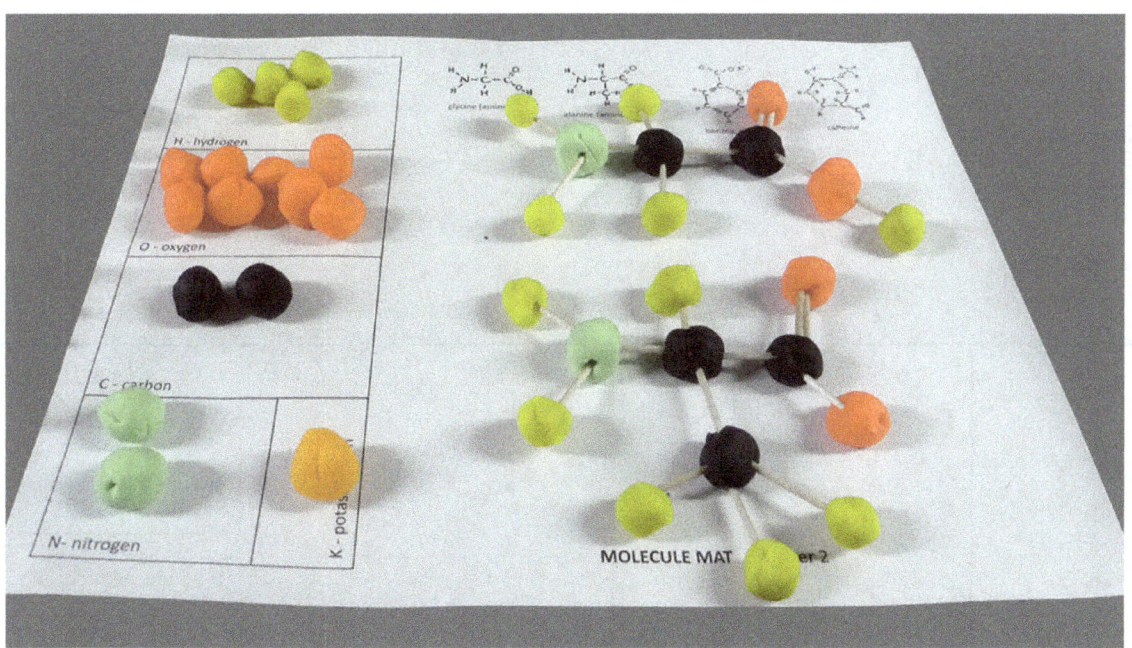

151

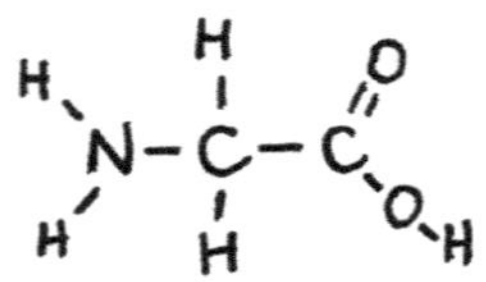

glycine (amino acid)

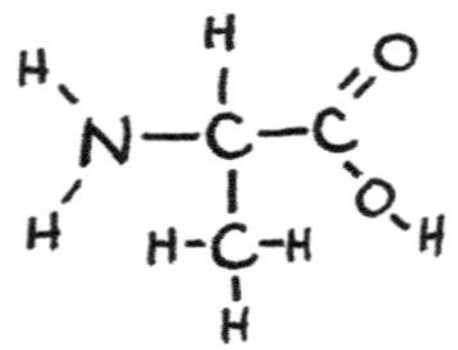

alanine (amino acid)

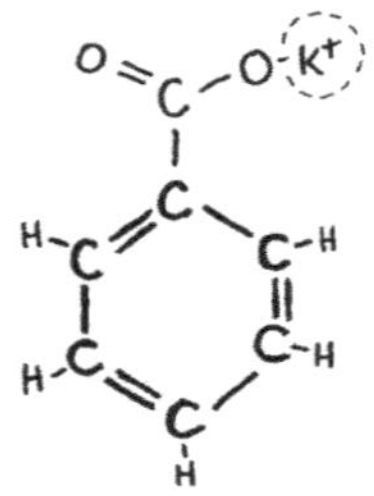

benzoic acid

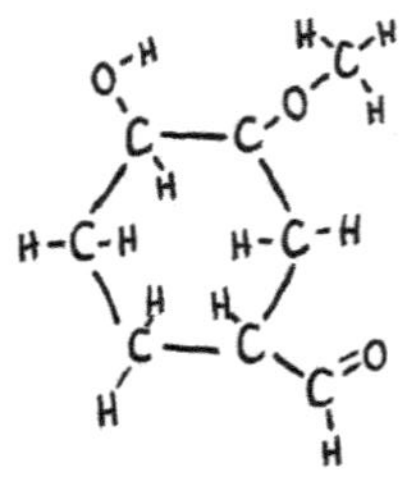

vanillin

H - hydrogen

O - oxygen

C - carbon

N - nitrogen

K - potassium

MOLECULE MAT Chapter 2

CHAPTER 3

ACTIVITY IDEA 3A   COOKING ACTIVITY: MAKE YOUR OWN BUTTER

You will need:
- a carton of heavy cream
- optional: a little salt to add to your cream
- a plastic or glass jar with screw-on lid
- something to go with your butter (bread, crackers, etc.)

What to tell the students:
    In this chapter you read about how cream is turned into butter by physically disrupting the fat globules and causing the triglycerides to be released. You don't need a fancy butter churn. You can make butter in a jar.

What to do:
1) Pour the cream into the jar, leaving about an inch (a few centimeters) of empty space at the top.
2) Add a tiny bit of salt (or you can have unsalted butter).
3) Screw the lid on very tightly.
4) Shake the jar until you see a solid form, with some liquid on top. This could take 10 to 15 minutes, so it is helpful to have several people share the work by passing the jar around.
5) Open the jar, pour off the "buttermilk" liquid, and then spoon out the fresh butter.

ACTIVITY IDEA 3B   COOKING ACTIVITY: MAKE YOUR OWN MOZZARELLA CHEESE (in less than an hour)

    There are several ways to make mozzarella. Some recipes call for liquid rennet, and others use vinegar. There are also a few different techniques for how to strain and/or stretch the curds. You should find a recipe that suits you and your situation. Just use the key words "how to make mozzarella cheese" in either an online search engine, or a video search engine.

ACTIVITY IDEA 3C   COOKING ACTIVITY: MAKE YOUR OWN COTTAGE CHEESE (in less than an hour)

    Cottage cheese is very similar to mozzarella. You use an acid (usually vinegar or lemon juice) to curdle the milk and separate the curds. For cottage cheese, the curds are patted dry, salted, then rubbed into crumbles. Some recipes call for cream to be added to the curds immediately before serving. Search online for a recipe that is a good fit for you. (There are also a lot of how-to videos on YouTube and other streaming services.)

ACTIVITY IDEA 3D   EXPERIMENT: MIXING MILK AND COLA

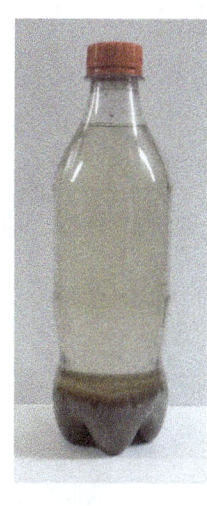

You will need:
- a personal-size bottle of cola (any brand, regular not diet)
- a tablespoon of milk

What to tell the students:
    In this experiment you will witness a reaction between the phosphoric acid in the cola and the proteins in the milk. The acid will curdle the milk, causing the proteins to clump together into heavy solids that will fall to the bottom of the bottle. The interesting part is that these curds will take all the artificial coloring down with them. You may need to let the bottle sit for a few days, but eventually most of the liquid in the bottle will turn clear, leaving a small brown layer at the bottom.

What to do:
1) Make sure your bottle of soda has not been shaken recently. In any case, open it very slowly, letting gas escape.
2) There will be a small amount of empty space at the top of the bottle. Pour in some milk until the bottle is almost full.
3) Screw the lid back on and gently tip the bottle back and forth, letting the milk mix with the cola.
4) Let the bottle sit undisturbed. You will start to see some results in a few hours, but for full effect let the bottle sit for several days.
5) If you need visual instructions, just search a video channel for "milk and coke experiment."

NOTE: The bottle will stay like this indefinitely. I have a bottle that has been sitting for three years.

ACTIVITY IDEA 3A   EXPERIMENT: USE IODINE TO TEST FOR STARCH

You will need:
- a small bottle of Lugol's iodine solution (TIP: This can be ordered from amazon.com or from homesciencetools.com.)
- eye droppers
- various specimens to test (Suggestions: cheese, milk, crackers, bread, cornstarch, rice, pasta, fruit, beans, carrots, etc. Also try a paper napkin, a piece of paper, a piece of paper towel, and a piece of newspaper.)

What to tell the students (while showing them the illustrations on the next page):
In this experiment you will use an iodine solution to test for starch. Iodine will cause starch to turn black or very dark brownish purple. Scientists are not entirely sure why this happens, but x-ray crystallography experiments suggest that iodine atoms form 3-atom molecules that are attracted to the inside of the coiled shape that amylose makes. Remember, amylose is made of long chains of glucose molecules. This chain coils into a helix shape. The iodine molecules find a cozy spot to sit right in the middle of the coils. The electrical properties of iodine are such that when they interact with the atoms in the amylose, they change the way it reflects light. The combination of iodine with amylose creates a situation in which almost all colors of light are absorbed. When an object absorbs all types of visible light, we see this as the color black.

NOTE: Lugol's solution is a combination of iodine and potassium iodide (KI) suspended in distilled water. This experiment doesn't work with just iodine. The addition of the potassium ions allows for the formation of these I_3 molecules.

What to do:
1) Lay out your test substances on a plate or paper towel.
2) Put a few drops of Lugol's solution on each specimen.
3) Observe results and discuss any results that were surprising.

ACTIVITY IDEA 3E   EXPERIMENT: HOW DO SALT AND SUGAR AFFECT YEAST GROWTH?

You will need:
- 6 packages of dried yeast
- warm water
- sugar
- salt
- lemon juice
- baking soda
- 6 small plastic disposable water bottles
- 6 balloons (Use larger balloons, the kind that can inflate to 12 inches.)

What to tell the students:
In this experiment, you will test the effect of salt, sugar, acids and bases on the growth of yeast. Yeast cells are microscopic, so you will not be able to see them unless you have a microscope. However, you will be able to observe the amount of carbon dioxide they give off as they eat and grow.

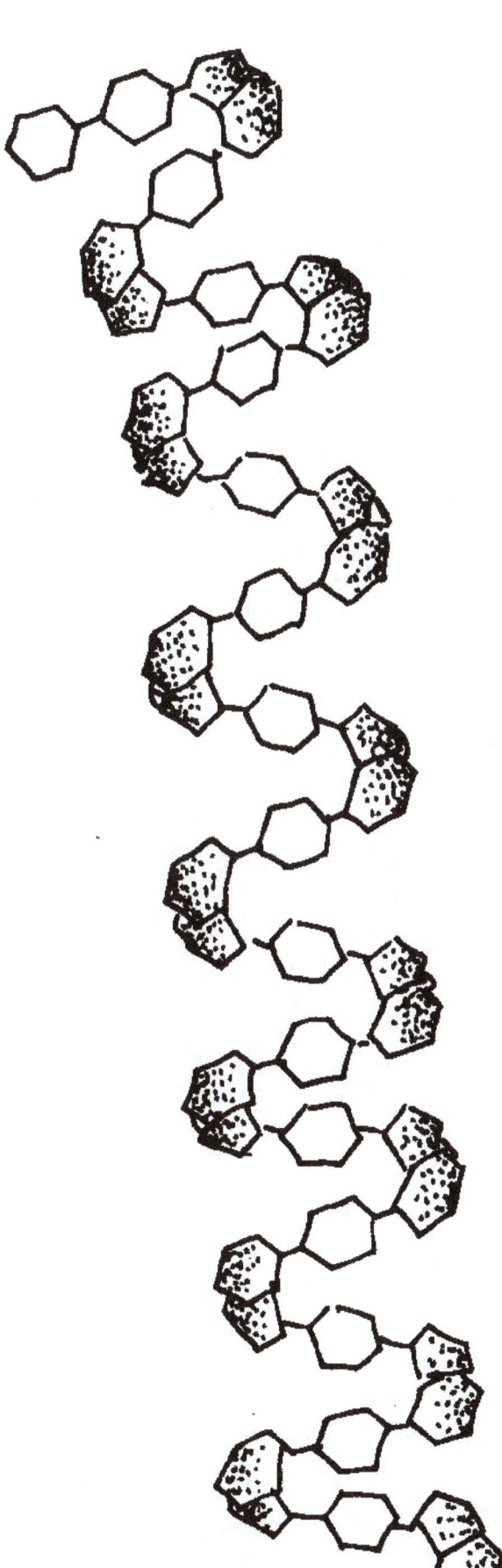

AMYLOSE
The hexagons are glucose molecules.

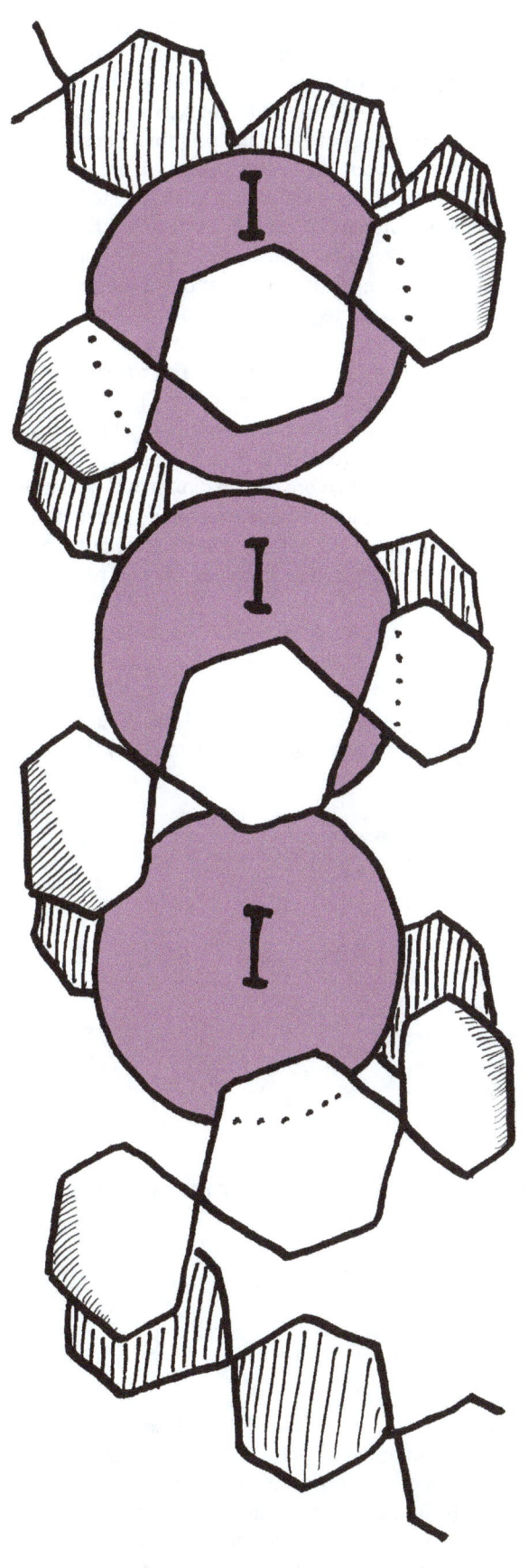

I = iodine atom
The iodine atoms form a 3-atom linear molecule that fits right into the coil.

What to do:
1) Take the labels off the bottles. Use a permanent marker to label the bottles 1 to 6, or put a written label on, or in front of, each bottle, so you know exactly what is in each.
2) Pour out half the water, or, if the bottles are empty, fill halfway with warm water.
3) Open the packages of yeast and pour one package into each bottle.
   Bottle 1 will be your "control." It will contain only yeast.
   Bottle 2 will have 1 teaspoon of sugar added to it. (The exact amount isn't important. If you are using metric, just use an approximate equivalent. What is important is to be consistent and use the same measurement for each bottle.)
   Bottle 3 will have 1 teaspoon sugar, and 2 teaspoons of vinegar.
   Bottle 4 will have 1 teaspoon sugar plus 1 teaspoon baking powder.
   Bottle 5 will have 1 teaspoon sugar, and 1 teaspoon salt.
   Bottle 6 will have 1 teaspoon sugar and 2 teaspoons salt.
4) Swirl the contents of the bottle just a bit to make sure everything gets mixed together. (You can even screw the lid on and shake the bottle, then take the lid off again.)
5) Stretch and flex the balloons a bit. You might even want to blow them up once then let the air out. Then put a balloon over the end of each tube (or bottle). The balloon will catch any carbon dioxide gas that the yeast will produce. The more the yeast grow and multiply, the more carbon dioxide will be produced.
6) Let the bottles sit undisturbed. You can go and do something else for a while and just check in from time to time. Don't disturb the balloons! The balloons need to stay fastened to the bottles for the duration of the experiment.
7) Let the experiment run until you see obvious results. One or two balloons should have a significant amount of gas in them.

EXPECTED RESULTS: Bottle number 2 should show the greatest amount of inflation, and bottle number 1 the least. Bottle 6 should show less activity than bottle 5 because it has more salt. The results for my bottles are shown below. Your results might differ from mine, due to a variety of factors, including type of yeast, temperature of water, type of vinegar, etc.
My results for bottles 3 and 4 surprised me. When I did this previously, I think I got some activity in my vinegar bottle.

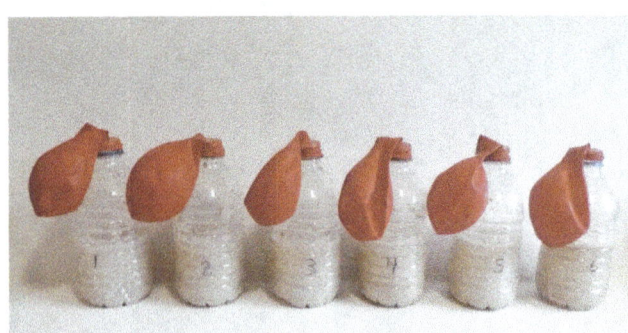

ACTIVITY IDEA 3F   MOLECULE MAT for chapter 3

You will need:
- a copy of the following pattern page for each student
- toothpicks
- materials for the atoms

What to do:
1) Put your chosen materials inside the boxes on the left side of the page (or in small dishes if they won't fit inside the boxes). Toothpicks can be set in a dish, or simply in a pile, within the students' reach.
2) Let the students work on their own as much as possible.
3) Build the four molecules shown, then recycle the parts to make Maillard molecules. There is no "right answer" to the Maillard molecules, since there are hundreds of ways the broken molecules could recombine.
4) For students who are keeping a portfolio of their work, take a photo of their paper with all the finished molecules on it.

NOTE: You can find full-color mini-posters about the Maillard reaction by doing an Internet image search for "Maillard reaction" or "Maillard reaction poster."

Make some Maillard molecules:
1) Use 4, 5 or 6 carbons.
2) Use 1, 2, or 3 oxygens.
3) Optional: add a sulfur or 1 or 2 nitrogens.
4) Fill any empty bonds with hydrogens. This means checking to see that all carbon atoms are bonded to four other atoms, all oxygens are bonded to two atoms, sulfur to two, and nitrogens to three. For example, if you find a carbon with only two atoms stuck to it, add 2 H's. If you find an oxygen with only one atom stuck to it, add one H.

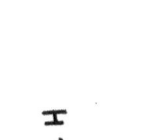

CH_3COOH acetic acid
(Vinegar is acetic acid + water.)

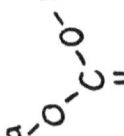

$NaHCO_3$
sodium bicarbonate

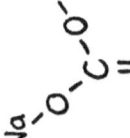

$C_3O_3H_3$ pyruvate

H - hydrogen
O - oxygen
C - carbon
S - sulfur
N - nitrogen	Na - sodium

MOLECULE MAT Chapter 3

ACTIVITY IDEA 3G   ACTIVITY: "GLUTEN WASHING TEST"

You will need:
- raw dough (You can use a homemade bread dough (even just flour and water is ok), or a purchased bread or pizza dough.)
- a sink and lots of water
- large bowl (several bowls if several students are participating)
- optional: strainer into which you can put the dough for final washing stages

NOTE: Students who have celiac disease or who have a true wheat allergy should not do this activity. They should probably not even be in the room while this activity is being done.

What to tell the students:
      You learned in this chapter that bread dough contains both starch and protein. In this activity you will wash the starch out of dough and end up with a lump of pure gluten protein. Starch is water soluble, while gluten is not. If you have ever made bread and had a hard time washing sticky bits of dough off your hands or out of your sponge, it's the gluten that you are having trouble with. This activity is called the "gluten washing test." Although it may seem simple, food scientists and professional bakers actually use this procedure to find out how much gluten is in a dough and to test the quality of the gluten.

What to do:
1) If you are using a homemade dough, knead it thoroughly to make sure the gluten is well developed.
2) Fill a large bowl with cool water and put the dough into it.
3) Knead and "massage" the dough while it is in the water. The water will turn cloudy.
4) When the water becomes very cloudy, pour it out and fill it with fresh water.
5) Repeat step 3, until the water is cloudy again.
6) Keep filling the bowl with fresh water and repeating step 3 until the water doesn't turn cloudy. At this point, the dough will be nothing but gluten and will tend to be tough and clumpy.

EXTENSION: Try this same activity with a gluten-free dough. Do you get different results?

ACTIVITY IDEA 3H   CRAFT: COLOR AND/OR DECORATE A GLUTEN STRING

You will need:
- copies of the following pattern page, one per student
- something with color to mark the highlighted amino acids (M, P, Q, C)

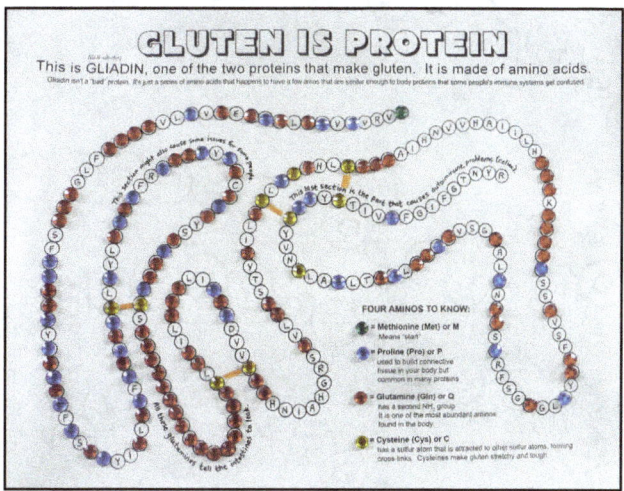

I happened to find some sticky-back "jewels" at my local dollar store, that were just about the size of the circles. I was able to find four different colors, so I had the students assign a color to each letter. They put a jewel to the left of each = sign (under FOUR AMINOS TO KNOW) and then went and found all of those letters in the long string and put the appropriate jewel on each. This made for a spectacular finished product!

HOWEVER, you can use anything you want to for the colors; even marker or colored pencil will be okay.

NOTE: I chose to highlight just these four amino acids so that the students won't feel overwhelmed with too much information. However, see the EXTENSION note below for advanced students who want to know more.

What to tell the students:
In this activity you will color code a piece of gliadin (GLIE-uh-din) protein, marking four types of amino acids. You will see which parts of this string cause problems in the body. Isn't it amazing that scientists have been able to make a map of this molecule? The page you will be working on would not have been possible without the efforts of many scientists working with high tech analysis machines.

What to do:
1) Have the students put dots of color (whatever you are using) next to each amino acids, to create a KEY. Make sure they read the notes on each amino acid.
2) Tell them to find these letters in the string and apply the appropriate color.
3) Tell them to find four places where two letter C's are directly across from each other and not far apart. Draw a line between them to indicate that this would be where a "bridge" would form, keeping those parts of the chain close together.
4) If they want to apply color to all the dots, make sure they are consistent and don't do random colors (make all the V's the same color, all the T's the same color, etc.).

EXTENSION: If the students are curious to know what the other letters stand for, use this list below. You might even give each student on amino acid to research before the next class and have each student present their findings and bring a diagram of what the molecule looks like.

AMINO ACID ONE-LETTER ABBREVIATIONS: (followed by their three-letter abbreviation)

A = alanine (Ala)
L = leucine (Leu)
F = phenylalanine (Phe)
D = aspartic acid (Asp)
H = histidine (His)
T = threonine (Thr)
N = asparagine (Asn)

G = glycine (Gly)
P = proline (Pro)
W = tryptophan (Trp)
E = glutamic acid (Glu)
K = lysine (Lys)
C = cysteine (Cys)
Q = glutamine (glutamic acid) (Gln)

I = isoleucine (Ile)
V = valine (Val)
Y = tyrosine (Tyr)
R = arginine (Arg)
S = serine (Ser)
M = methionine (Met)

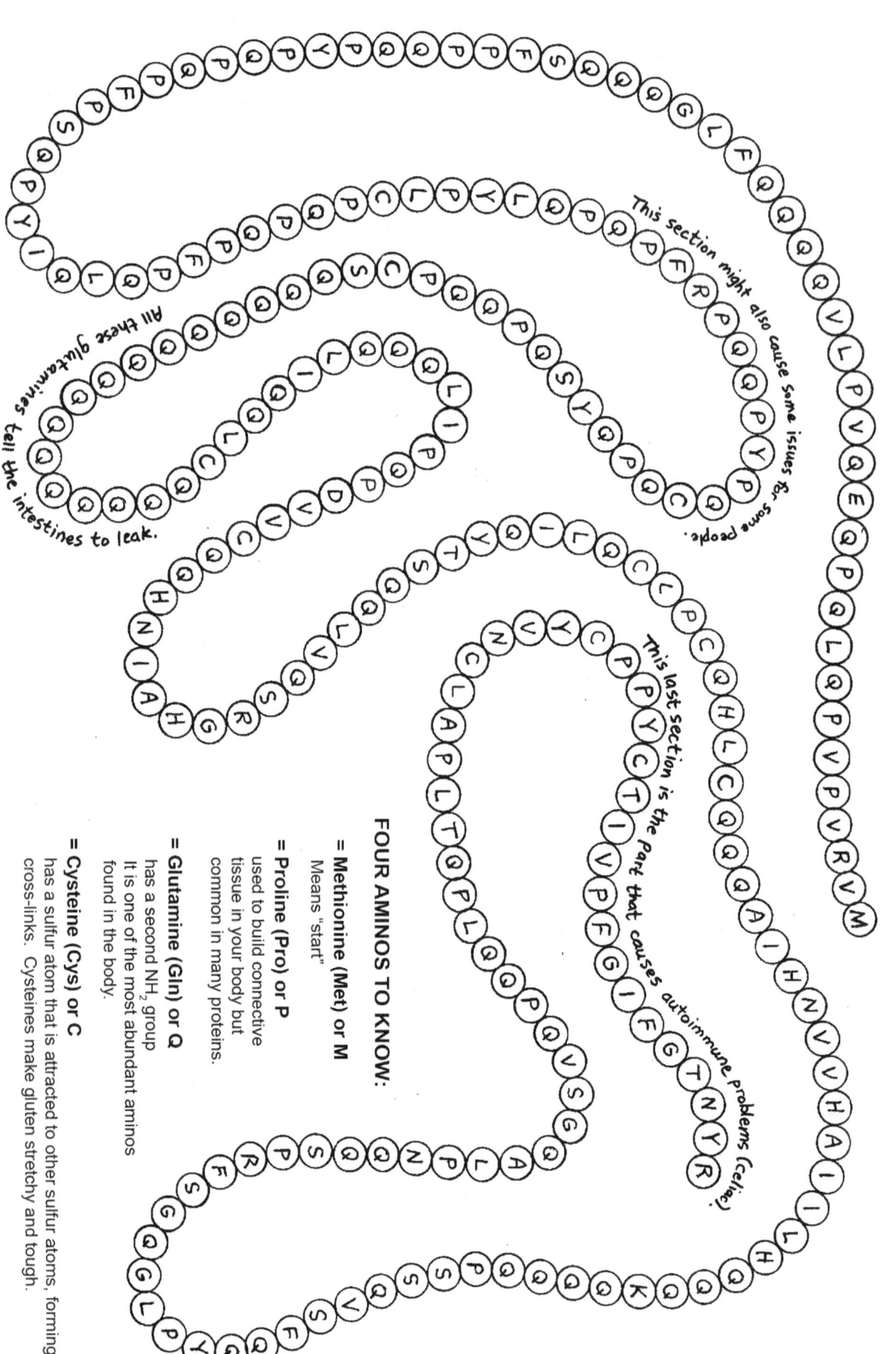

ACTIVITY IDEA 3I   GAME: "THE ENZYME GAME"

This game is also a free download on my website. It is printed here for your convenience.
The game works best with two to four players, but in pinch, six can play. Four players can play in about 20 minutes. Two players will take more like 30 minutes. Allow 30-45 minutes to make the game.

You will need:
- copies of the following pattern page printed onto card stock (If you need a digital copy, just download from my site: www.ellenjmchenry.com, click on FREE DOWNLOADS, then click on HUMAN BODY and look for "The Enzyme Game" under the Nutrition category.)
- paper fasteners
- small washers (though you can probably make the spinners work without them)
- scissors and glue
- a few sheets of thin cardboard (cereal box cardboard will work)

paper fastener

How to set up:
1) Copy the pattern pages onto card stock. Print two copies of each card page if you will be playing with 2 to 4 players. Make 3 copies of each page if you will be playing with 5 or 6 players.
2) Glue the spinner patterns onto thin cardboard. If you are using cereal boxes, glue to the blank side. The printed side is too shiny and the glue may not adhere well. Press and flatten and allow to dry for at least 10 minutes. Longer is better, but I have found that I can move to the next step with even a short dry time.
3) Cut out the spinner squares and the circles and arrows. Punch holes at the center dots. I start the holes with the point of a scissor or compass, then use a ballpoint pen to make them larger. Don't make them larger than the head of the paper fastener.
4) Assemble the spinners as shown on the pattern sheet. Use a washer between the spinner at its base. This will allow the spinner to turn more freely. Don't over tighten the paper fastener. You may have to fuss with the spinner a bit to get it to spin well. Just keep the tension loose and then spin it a number of times to "break it in" before using it with the kids.
5) Cut apart the cards. OPTION: You can add extra notes on the cards if you think your players will need extra info. For example, you might want to add on the Fructose card, "Came from glucose." These notes are not necessary, as you can figure this out as you go along, but if you want your players to have extra hints, feel free to add them.

How to play:
The rules of this game are somewhat flexible. This is not to frustrate you or to make things confusing, but rather to allow you to tailor the game to the age, ability, and personality of your players. When I played it with my class, I found that each group had a different idea about what "trade" should mean. I allowed them to determine how to manage their trading as long as everyone agreed to the same rules. This worked out very well and made the game much more cooperative than I had expected. I design my games to have as much interaction between players as possible, and little or no "dead" time in which players have to wait.

1) The object of the game is to get as many complete sets of three as possible. A complete set consists of an enzyme, the substance on which it operates, and the broken down parts. Example: Starch, Amylase, Glucose. (However, you can't lay down a set until you have correct pH and temperature.)
2) Consider laying out the cards before you start the game and having the players find the sets. This way they will know what they are looking for as the play the game.
NOTE ABOUT GLUCOSE ISOMERASE: The two cards that go with this enzyme are glucose and fructose (separate cards).
3) Shuffle the cards and then place them face down not in a stack but in a loose pile so players may choose any card they wish. (This will make discarding easier, also.)
4) Players start with 3 cards. They should have at least 3 cards in their hand at all times. If, after laying down a set of cards, they end up with less than 3, they should draw enough cards so they have 3.
5) All players will move at once. You don't have to take turns in this game. This avoids "dead time" and makes the game a constant buzz of activity, which I have found is the best possible scenario for optimum use of group time. The opening of each "move" is to spin the large spinner. Players can take turns spinning. If it lands on "Draw a Card" everyone draws a card. If it lands on "Discard" then each player choses one of their cards to return to the draw pile. (This allows for more shuffling of the cards during the game, allowing players to get rid of cards they can't use.)
6) Here's the most flexible part of the game—when the arrow lands on "Spin or Trade."
If the "Spin" option is chosen, the player spins both the temperature and the pH spinners. If they happen to land on the required temperature and pH for their set of three cards, they may then retire that set, turning it face down in a "keeper" pile for all the completed sets. Choosing "Trade" means they trade a card with another player. However, you've got some options here. Probably the best way to explain these options is to start with an anecdote.

When I designed the game, I planned that if any player in the game had a set of three they wanted to lay down, they could call out, "Spin," and therefore that turn was devoted to "Spin" and no one could trade. The pH and temperature spinner would be spun. Then, anyone who had a set of three for which those pH's and temp's were good, could then lay it down (put sets face down to avoid any confusion with other cards they might lay on the table). Then, I thought that if no one had a set of three, the "Trade" option would be the default option, and this would mean that each player must choose one of their cards and give it to the player on their right.

Here' what happened in my class: The players wanted to be able to choose for themselves which option they wanted, trade or spin. Those who had sets ready to lay down each got a turn to try to spin the correct pH and temp. (If either temp or pH wasn't suitable, they had to wait and try again next time the arrow landed on "Spin or Trade.") Those who wanted to use the "Trade" option laid out their cards on the table so everyone could see them, then they began to barter for trades. They could see cards they needed and they tried to arrange suitable barters that the other players would agree to. This actually made the players look very closely at not only their cards but also other players' cards and gave them a greater awareness of all the cards. I had not thought of this beforehand. So I allowed this change to the rules. I told the teams that the trades had to be mutually agreeable. If someone refused to trade, then there was no trade. They all agreed. However, in order to persuade the other player to give them a card, they started offering multiple cards in order to get the one card they needed. Another thing I had not predicted! So we had one group that operated their own little "economy," excitedly arranging elaborate trade deals with each other. They had a lot of fun and paid great attention to what cards they had. It worked very nicely and they all had a lot of fun.

I always tell my students that as long as all players agrees to the rules and everyone plays by those rules, then the game is fair.

So you need to decide what rules you will use for the "Spin or Trade" option. If you want to make it a "blind" trade where each player hands a card to the player on their right, then have them keep their hands a secret from the other players. If you want to use the option where players are allowed to barter, they will need to lay out their cards so everyone can see them. You know your players and can choose the option that is best for your situation.

If a situation comes up for which a rule is not given here, just supervise the team as they decide on how to make a rule to cover that situation. If there is disagreement, the adult supervisor gets to decide what the new rule will be.

Remember, the whole point of the game is to reinforce leaning about enzymes, not to get sets of 3 and win. To increase learning (especially with players in the 10-13 age bracket), the adult supervisor can sit with the team and add comments at appropriate points. For example, I had opportunities to help players by getting them to look closely at the pictures, asking them what the hexagons and pentagons represented. I also refreshed their memory about the difference between cellulose and starch and told them to look carefully at the glucose "flags" to see which way they are oriented. This gave me the opportunity to do tiny bits of individual tutoring at various points.

7) The game is over when all the cards in the draw pile are gone. Whoever has the greatest number of completed sets on the table (the ones they spun correct pH and temp for) wins. Sets still in their hands can be used as tie-breakers.

OTHER GAMES YOU CAN PLAY WITH THESE CARDS:

1) You can play a standard "Go Fish" type game where players try to get sets of three by either asking other players for cards or drawing from the draw pile.

2) A standard "memory" card game where the cards and laid out and players turn cards over trying to get matches. In this case, players would get to turn over three cards on each turn instead of just two.

SUCRASE

Breaks apart: **sucrose**
Works best at **pH 6**
Most efficient temp: **37.5° C**

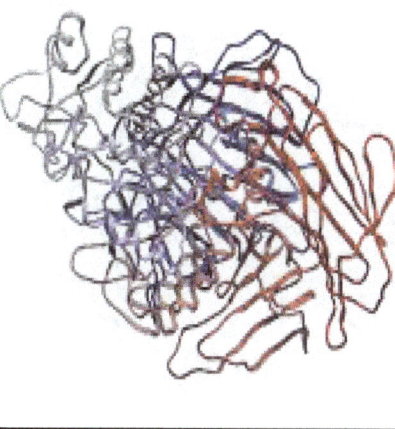

PEPSIN

Breaks apart: **proteins**
Works best at **pH 1-3**
Most efficient temp: **37.5° C**

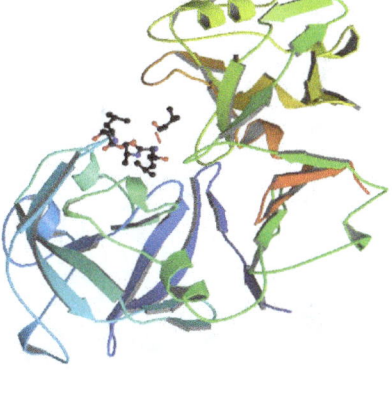

AMYLASE

Breaks apart: **starch**
Works best at **pH 6-7**
Most efficient temp: **37.5° C**

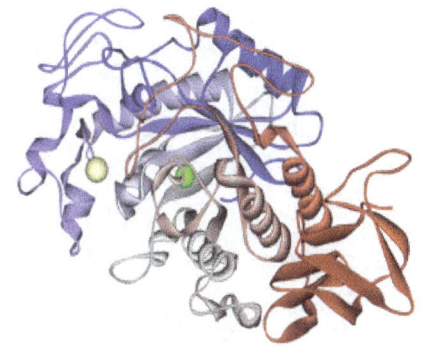

GLUCOSE ISOMERASE

Changes glucose to fructose
Works best at **pH 7-8**
Most efficient temp: **60-70° C**

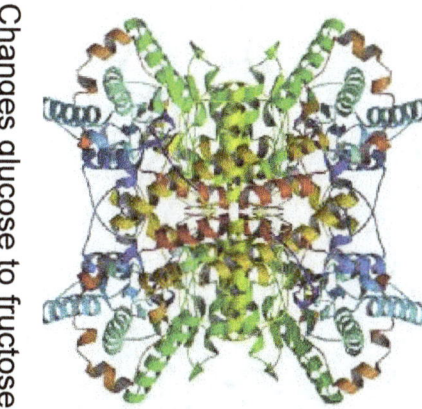

LIPASE
(Pancreatic lipase)

Breaks apart: **triglycerides**
Works best at **pH 8**
Most efficient temp: **37.5° C**

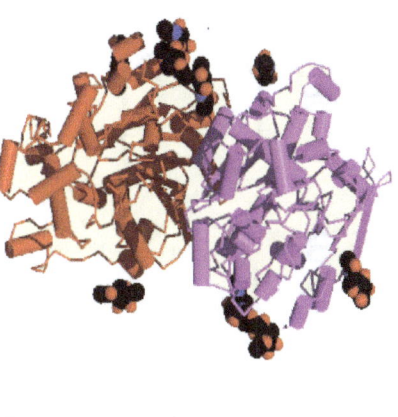

LACTASE

Breaks apart: **lactose**
Works best at **pH 8**
Most efficient temp: **37.5° C**

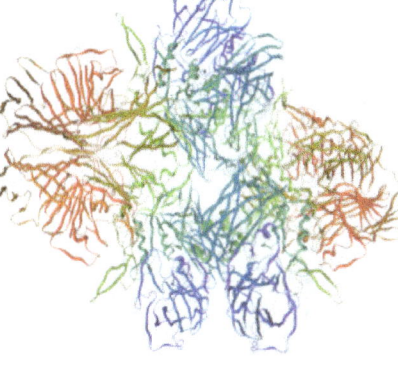

PECTINASE

Breaks apart: **pectin**
Works best at **pH 4-5**
Most efficient temp: **30-40° C**

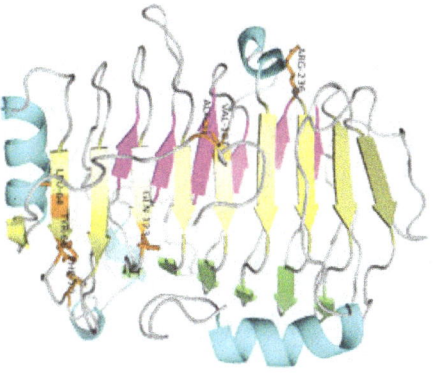

CELLULASE

Breaks apart: **cellulose**
Works best at **pH 5**
Most efficient temp: **40-50° C**

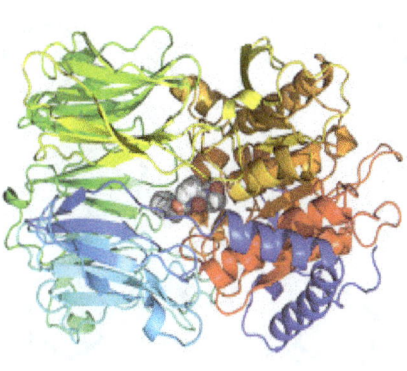

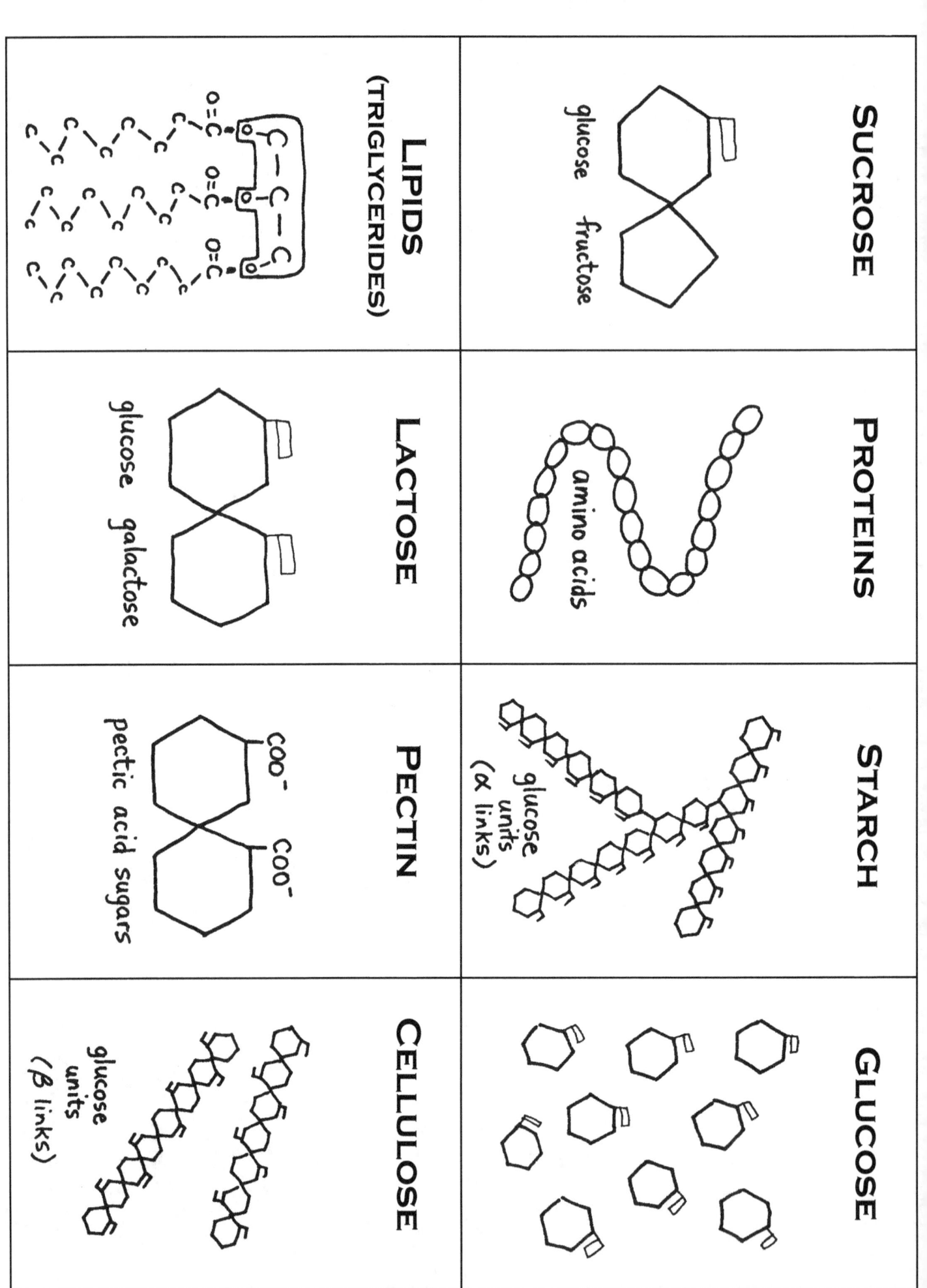

GLUCOSE & FRUCTOSE	GLUCOSE & FATTY ACIDS
AMINO ACIDS	GLUCOSE & GALACTOSE
GLUCOSE	PECTIC ACID SUGARS
FRUCTOSE	GLUCOSE

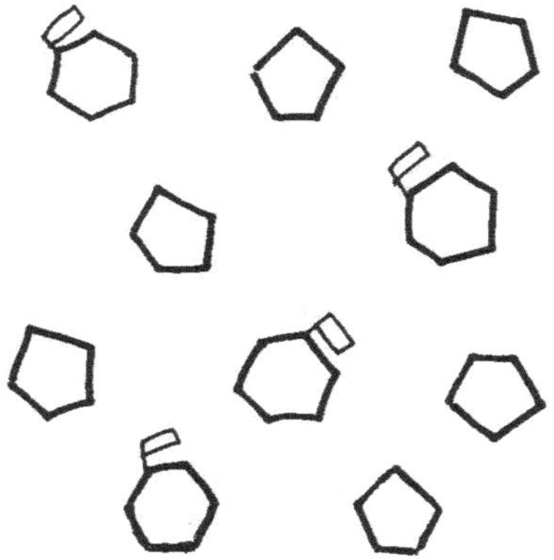

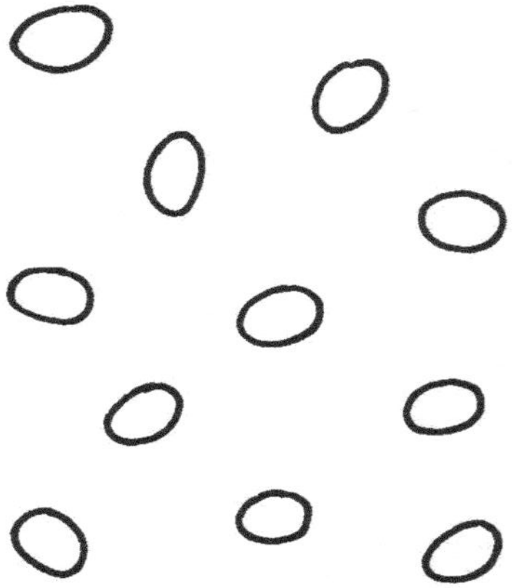

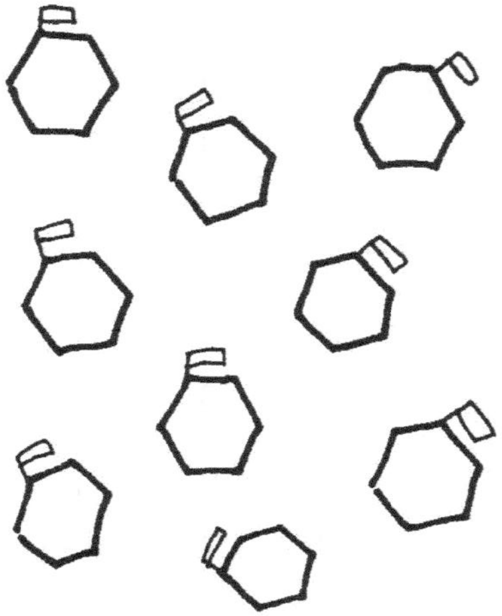

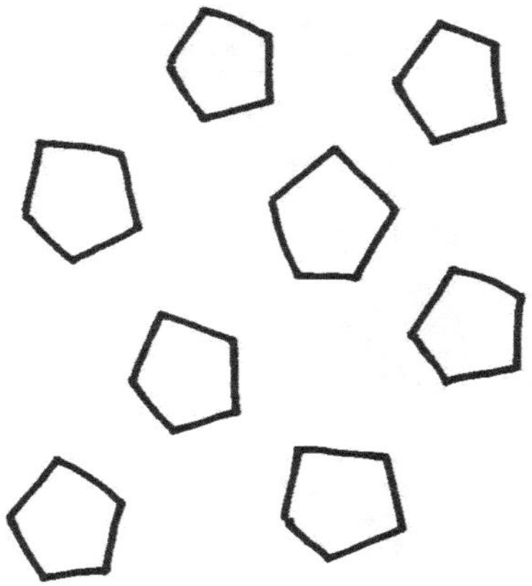

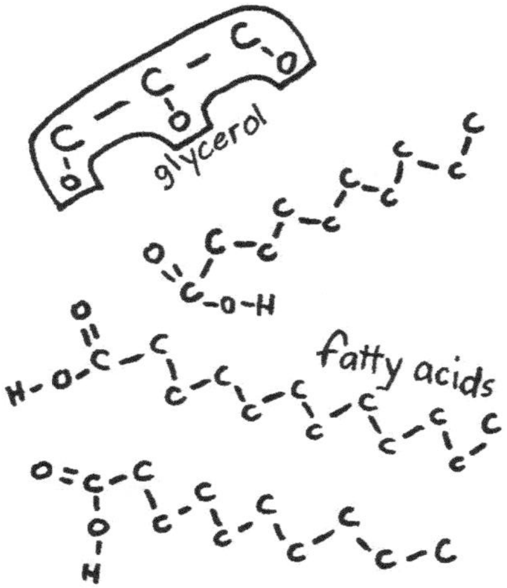

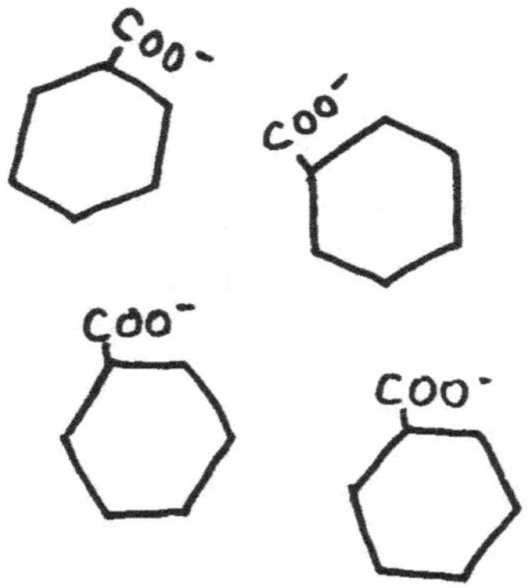

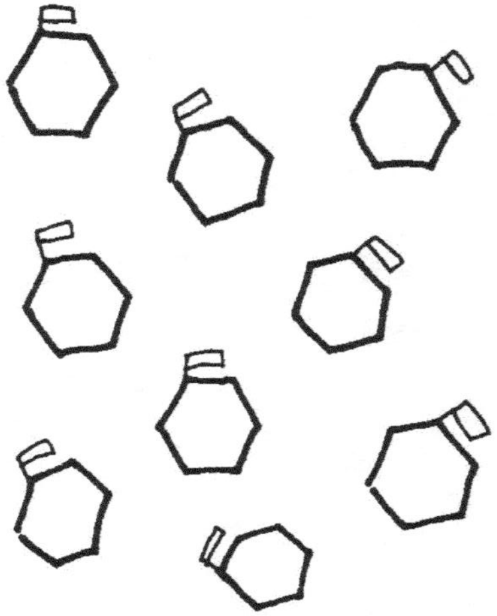

PRINT 2 COPIES OF THIS PAGE IF YOU ARE PLAYING THE GAME WITH 2-4 PLAYERS. FOR 6 PLAYERS, MAKE 3 COPIES.

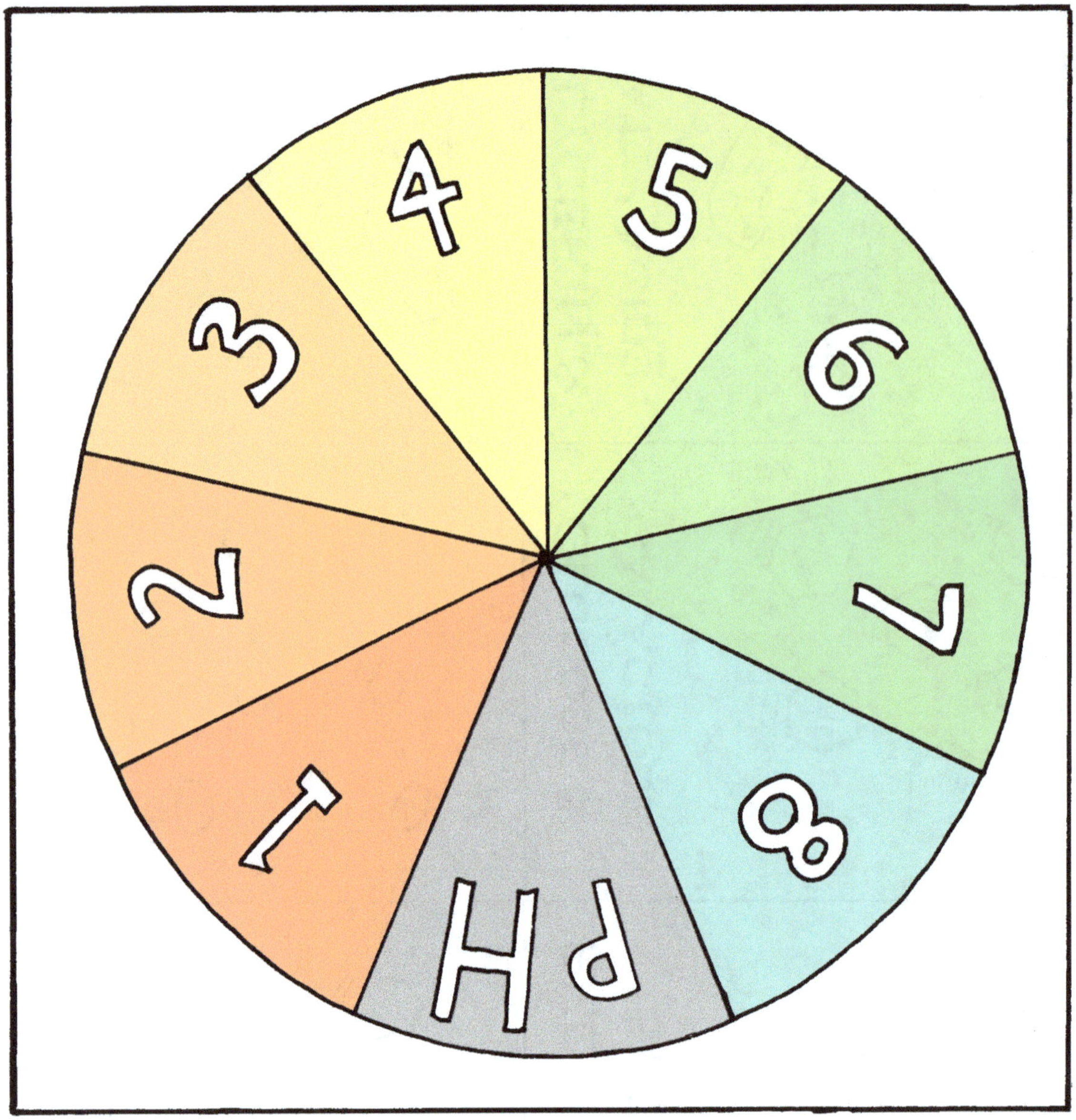

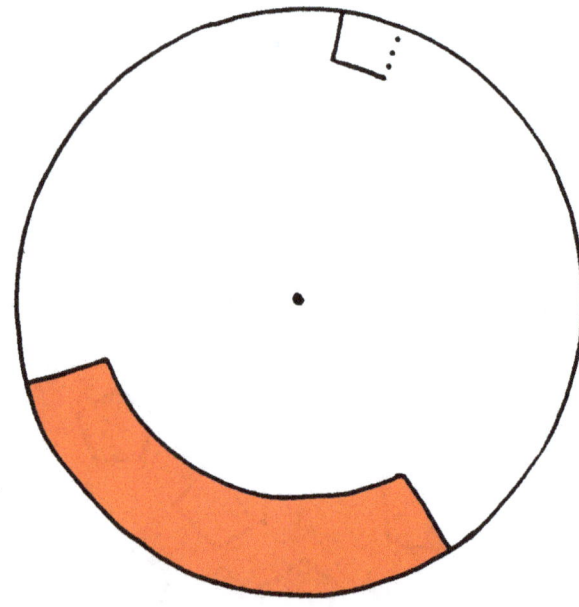

The red bar replaces a pointer. Any wedge (even a small portion of it) that touches the red bar is okay.

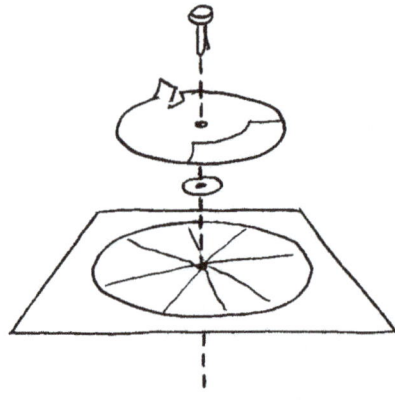

NOTE: Cut the notch on the wheel and fold it up to make a tab that can be used to flick the spinner.

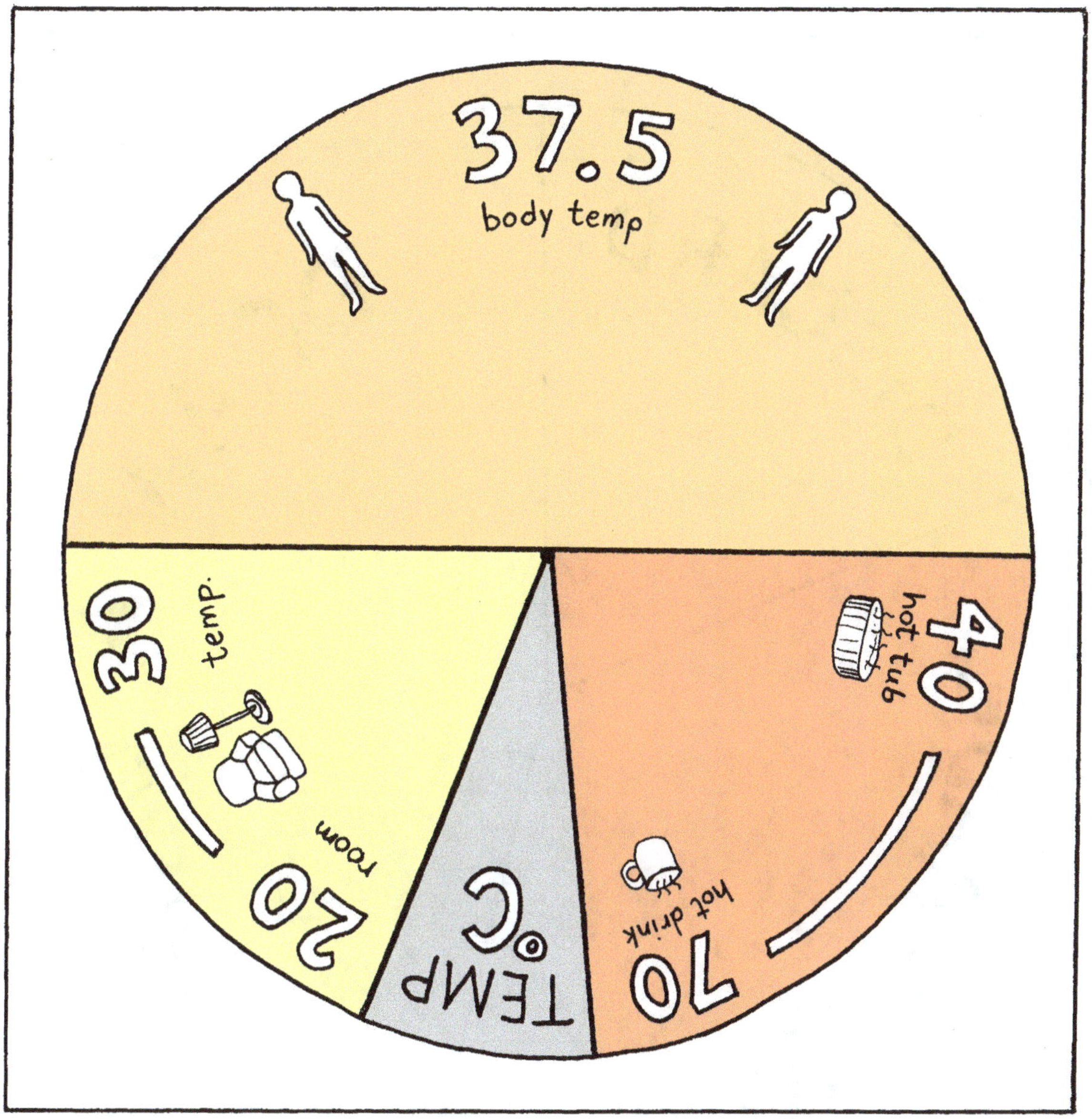

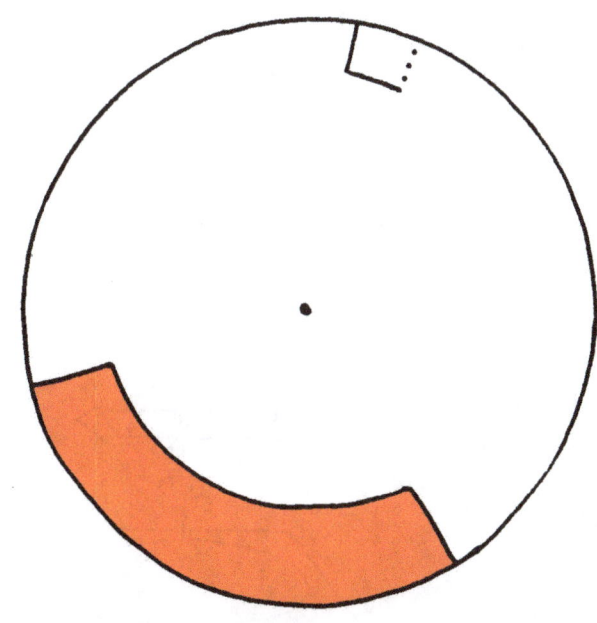

The red bar replaces a pointer. Any wedge (even a small portion of it) that touches the red bar is okay.

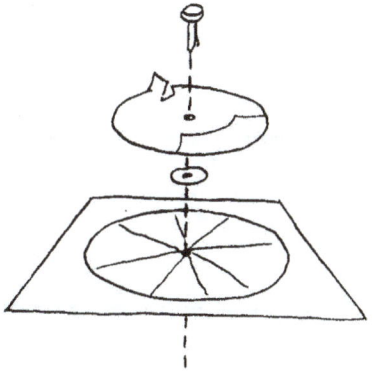

NOTE: Cut the notch on the wheel and fold it up to make a tab that can be used to flick the spinner.

167

DISCARD
DRAW A CARD
SPIN or TRADE

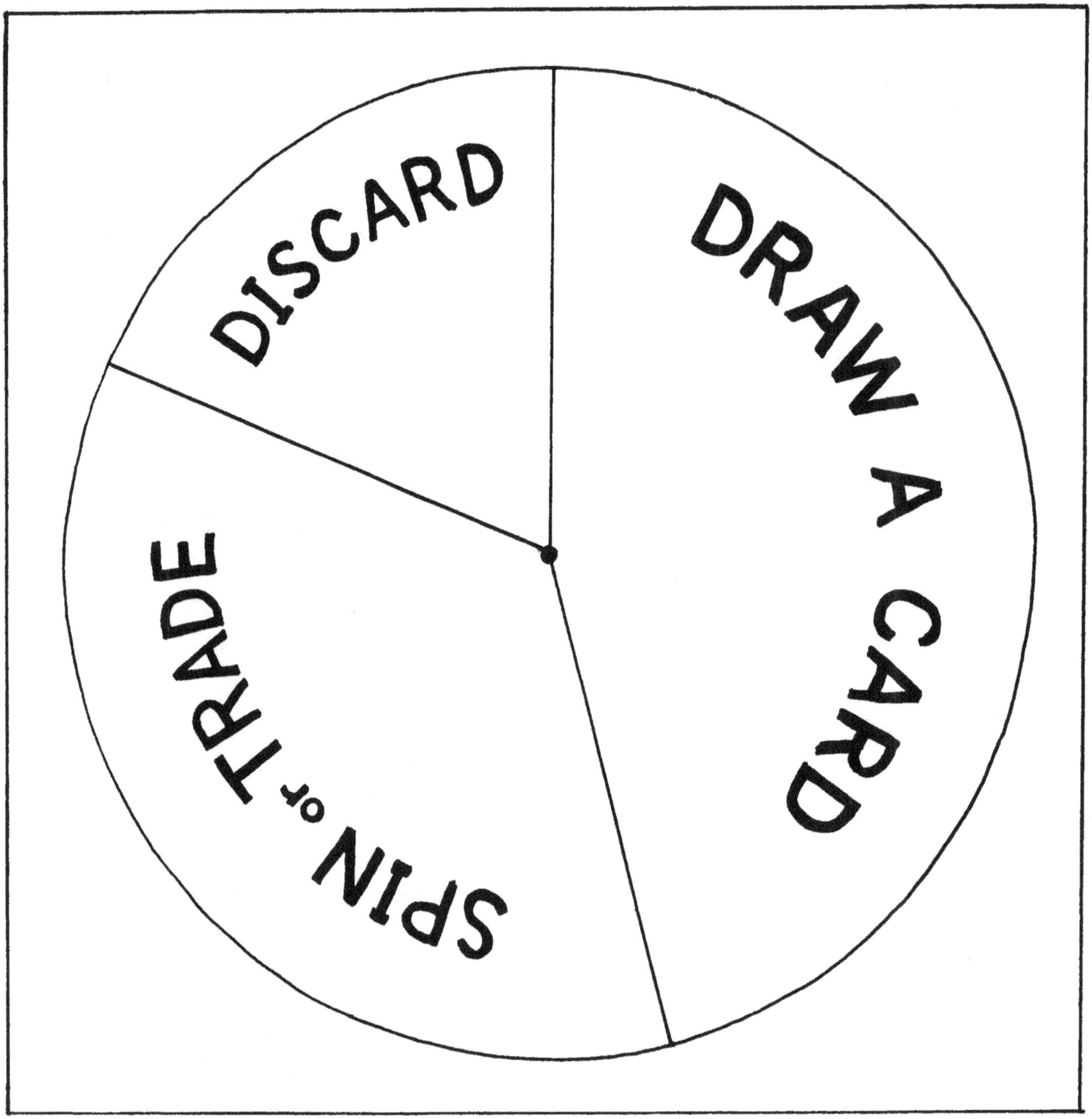

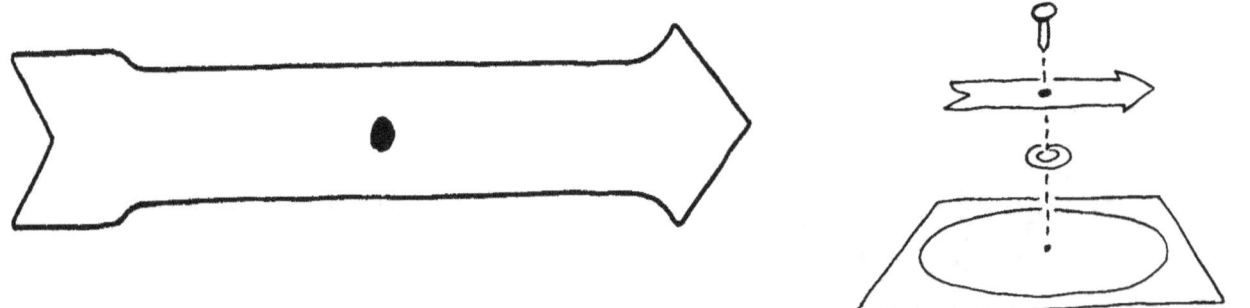

CHAPTER 4

ACTIVITY IDEA 4A   EXPERIMENT: TESTING BANANAS FOR STARCH   (See idea 5F, which also uses bananas.)

You will need:
- four (or more) bananas, ranging from very green to very ripe
- Lugol's solution (iodine)
- eye dropper

What to tell the students:
  In this experiment we will see how the chemical composition of a banana changes as it becomes ripe. You know that very ripe bananas are much sweeter than green ones. This test will prove that starch is converted to sugar as a banana ripens. Can you name the agent of this change?

What to do:
1) Cut a slice of each type of banana.
2) Put a few drops of Lugol's solution on each slice.
3) Observe any darkening. Iodine stains starch molecules but not sugar molecules. Which banana has the most starch?
4) Can you think of an explanation for the chemical change that is going on? What is starch made of? What type of molecule is able to tear starch molecules apart?

EXPECTED RESULT: The green banana will stain black, or very dark. The riper the banana, the less darkening you will see. This is because plants always contain the enzymes needed to break down the starches in their fruits. The fruit is supposed to be nourishment for the seeds inside the fruit. The sugars can provide the growing plant with energy. The bananas must contain amylase or a very similar enzyme that is able to break the starch into individual glucose molecules.

ACTIVITY IDEA 4B   EXPERIMENT: CAN WE PREVENT APPLE SLICES FROM TURNING BROWN?

This activity requires only a short time to get started, then it will need to sit for a few hours.

You will need:
- an apple
- small bowl of water
- small bowl of lemon juice
- small bowl of vinegar
- small bowl of salt water
- small bowl of milk
- tongs
- paper towels or paper napkins
- marker or pen

What to tell the students:
  We've all seen apple slices go brown. This process is called "enzymatic browning." The enzyme that does this is called "polyphenol oxidase." When this enzyme combines with oxygen in the air, it creates a brown molecule called melanin. In this experiment we will find out what type of substance can prevent, or slow down, the browning process.

What to do:
1) Cut 6 pieces of paper towel or napkin and write one of these words on each: CONTROL, WATER, LEMON JUICE, VINEGAR, SALT WATER, MILK
2) Cut the apple into slices.
3) Put a slice on the paper that says CONTROL. This will let us see what happens if a slice is not dipped into a solution.
4) Use the tongs to dip an apple slice into the water. Hold it under the water for about 30 seconds. Then pull it out and set it on the paper that says WATER.
5) Use the tongs to dip an apple slice into the lemon juice; hold for 30 seconds, then place on the paper that says LEMON JUICE.
6) Rinse the tongs, then use them to dip an apple slice into the VINEGAR. Place onto the labeled paper.
7) Use this process for the rest of the solutions.
8) Let the slices sit for at least an hour, then observe. Feel the slices. Are the textures different?

9) Let the slices sit for a few more hours, then observe again.
10) Discuss the results. Which solution prevented browning? Was there a difference between the slice that was not dipped at all and the slice that was dipped in water? What did vinegar do? Is the milk-dipped slice harder than the rest? What happened to the salt water slice? Did any of your results surprise you?

EXPECTED RESULT: The undipped slice will turn brown. The water-dipped slice will show some browning but not as much as the control slice. The slice dipped in lemon juice will show much less browning. (The action of the polyphenol oxidase enzyme is blocked by acid.) Some people who have done this experiment reported that their salt water slice had less browning than their lemon juice slice. It has also been reported that the milk slice felt harder than the others, and that the vinegar slice turned mushy. Another possible surprise is that the vinegar slice may turn brown even though vinegar is acidic like lemon juice is.

Your results might or might not match these reported results, but you should definitely see less browning on the lemon juice slice than the control.

INTERESTING FACT: Commercial apple growers are able to keep apples in good condition for a long time by putting them in a special cooler that not only keeps the temperature low but also takes the oxygen out of the air so the apples are in a pure nitrogen atmosphere. (Normal air is about 80 percent nitrogen, 20 percent oxygen.)

ACTIVITY IDEA 4C   FRUIT AND VEGETABLE CARD GAMES

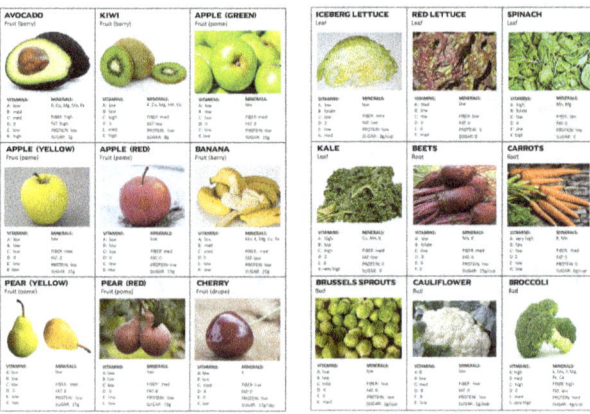

This activity requires you to access the pattern pages via my website. (Go to www.ellenjmchenry.com, click on FREE DOWNLOADS, then click on PLANTS.) This is necessary because of possible copyright issues. The photographs on the cards are from a wide variety of sources and are fine for educational use on a free download, but could potentially create issues in a copyrighted book that is being sold for profit.

As you can see, the cards are high quality and are definitely worth printing on heavy card stock paper. You can use them for reference, not just for games. There are 6 other pages, besides the ones shown here, for a total of 8 pages.

The download also has suggestions for games you can play with the cards. **Choose one or two activities to do with the cards after this chapter, and save a few for the next chapter. This activity will be listed again for chapter 5.**

ACTIVITY IDEA 4D   SNACK: MAKE HOMEMADE MAYONNAISE (You can easily find alternative recipes online.)

You will need:
- one large egg (You will use it raw. If you are worried about safety, buy a pasteurized egg.)
- 1 tablespoon Dijon mustard
- 1 tablespoon vinegar
- 1/4 teaspoon salt
- 1 cup vegetable oil (don't use extra virgin olive oil)
- optional: 1 teaspoon lemon juice
- a food processor or stick blender (something with blades that go really fast)

What to do:
1) Put egg in food processor, or in bowl if you are using a stick blender. Process for 20-30 seconds.
2) Add mustard, vinegar and salt. Process for another 20-30 seconds.
3) Begin to add oil slowly, a few drops at a time. This is necessary for the emulsion to happen. The mixture should begin to thicken at this point.
4) After about 1/4 of the oil has been added, you can begin to add it a little faster. Scrape the sides and bottom of the bowl to be sure all ingredients have been blended.
5) Optional: When mixture is thick, you can add some lemon juice or some more vinegar to adjust the flavor.
6) If the mixture seems too thin, turn the processor or stick blender back on, and slowly add a tiny bit more oil until it thickens.

ACTIVITY IDEA 4E   EXPERIMENT: USE ANTHOCYANIN AS pH INDICATOR

This is a classic science activity, so you, or some of your students, may have worked with cabbage water already. If so, try to find some unusual substances to test—substances that are unlikely to have been included in any previous experiments.

You will need:
- a head of purple cabbage
- a pot of water in which to boil the cabbage leaves
- eye droppers
- copy (or copies) of the following pattern page with the numbered colored dots
- a variety of substances to test

SUGGESTIONS (choose the ones that are best for your situation): water. salt water, sugar water, vinegar, lemon juice, apple juice, orange juice, baking soda in water, cola, seltzer water, flavored seltzer (example: LeCroix® brand), milk, laundry detergent, shampoo, mouthwash, hand soap, toothpaste, window cleaning fluid, ammonia, bleach. (If you want an extremely alkaline substance, use liquid drain cleaner, but be sure to supervise it carefully as it can cause injury to skin. Perhaps give them only a few drops to test.) **Make sure to include water** (distilled if possible) as a neutral 7.0 to which the other samples can be compared.
- You will also need one of the following set-ups for your substances. Choose the option that works best for your situation.

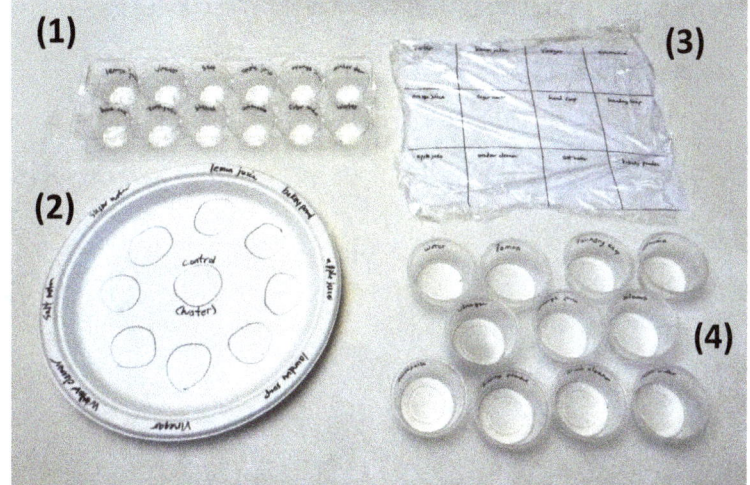

1) Use a foam or clear plastic egg carton, which gives you 12 cups for 12 substances. The downside with this method is that it is hard to label the compartments. You could tape little paper slips to the compartments, or you could use a permanent marker to write on the inside of the top of each compartment.

2) Use a plastic or foam plate and write the labels around the outside of the plate (where the numbers on a clock would be) and then use a dropper to squeeze out a small puddle of the substance close to its label. Put the water (control) in the center.

3) Write the labels on a sheet of paper then put a piece of plastic wrap on top of the paper, and squeeze out small puddle of each substance right near each label.

4) If you want the students to be able to arrange all their substances in order, from acidic to basic, you will need to provide individual small cups for each substance. Clear plastic condiment cups are about the right size. Or, you could cut apart the compartments of foam or plastic egg cartons. Disposable 4-oz plastic (bathroom size) cups also work.

Preparation ahead of time:
1) Chop your purple cabbage (if it is a large head you can use just half of it) and put it into a pot with enough water to cover the cabbage. Boil for a few minutes, until the water in the pot is dark purple. Turn off the heat and let it cool to room temperature. Pour off the purple liquid and save it in a jar or container with a lid. It will keep in the refrigerator for a few days if necessary.
2) Make a copy (or copies) of the pattern page with all the colored dots. Cut it into strips so each student can have a numbered strip. They will compare their colored solutions to the strip to determine the solution's number on the pH scale.
3) Round up the materials for the option you chose (1 - 4) above.

What to tell the students:
In this experiment you will see how anthocyanin pigment taken from purple cabbage can be used as a pH indicator to test how acidic or alkaline a substance is. (Remember that the word "alkaline" means the same as "base" or "basic.") Anthocyanin will turn pink or red in the presence of an acid. (Think of sour, unripe raspberries, which are red.) In a neutral environment (like a ripe berry) anthocyanin will be blue or purple. In an alkaline (basic) environment, anthocyanin will turn green or greenish-yellow. You will be given a strip of paper with colored dots that display the range of colors you may see. Each dot has a number that represents a number on the pH scale. With this color scale you can figure out an approximate pH number for each substance you test.

IN CLASS:
1) Give each student, or pair of students, a set-up for testing. Make sure they understand what to do **before** any substances are given to them.
2) Distribute the samples of substances to be tested. (WAIT to give them purple cabbage water until all sample substances are on their plate.) This might require several adults with bottles and eye droppers going around and putting the test substance right next to its label. If you don't have enough adult help to do this, the other option is to send around labeled containers with a dropper in each one, and tell the student they are to squeeze out a small sample of each, putting it in the proper place, and to make sure to keep each dropper with its container. If the eye droppers get mixed up, your samples will be contaminated and this will effect your results.
3) Give each student, or pair of students, a cup of purple cabbage water and an eye dropper. Students should add a dropper of purple anthocyanin water to each substance and see what happens to the color.

NOTE: There isn't an exact amount of purple solution that must be added. Tell them to add drops slowly and just watch until the color stays about the same even after adding a few more drops. When the color is stable, stop adding drops. Fortunately, there is a wide tolerance for error. The experiment will still work even if they add too much, or too little, purple cabbage solution.

Depending on the substances you test, you really can get a wide range of colors. Here is an actual example of student work in one of our classes. Each puddle has anthocyanin indicator in it, but as you can see, the pH of the substance has drastically altered the original purple color in most cases. (We put our papers inside a plastic sheet protector. It worked really well!)

Example of student work:

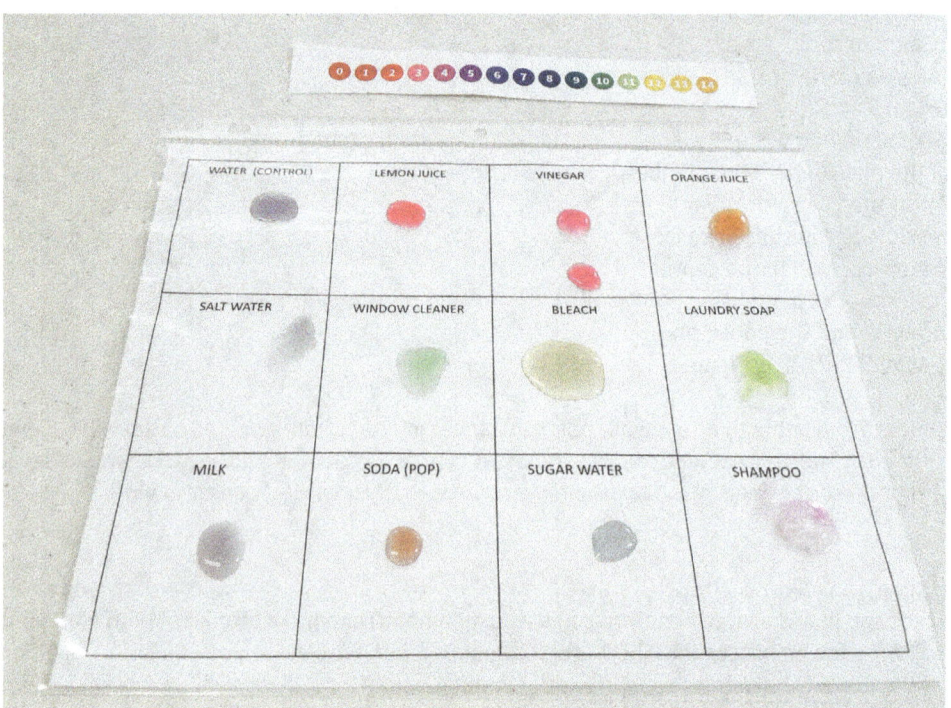

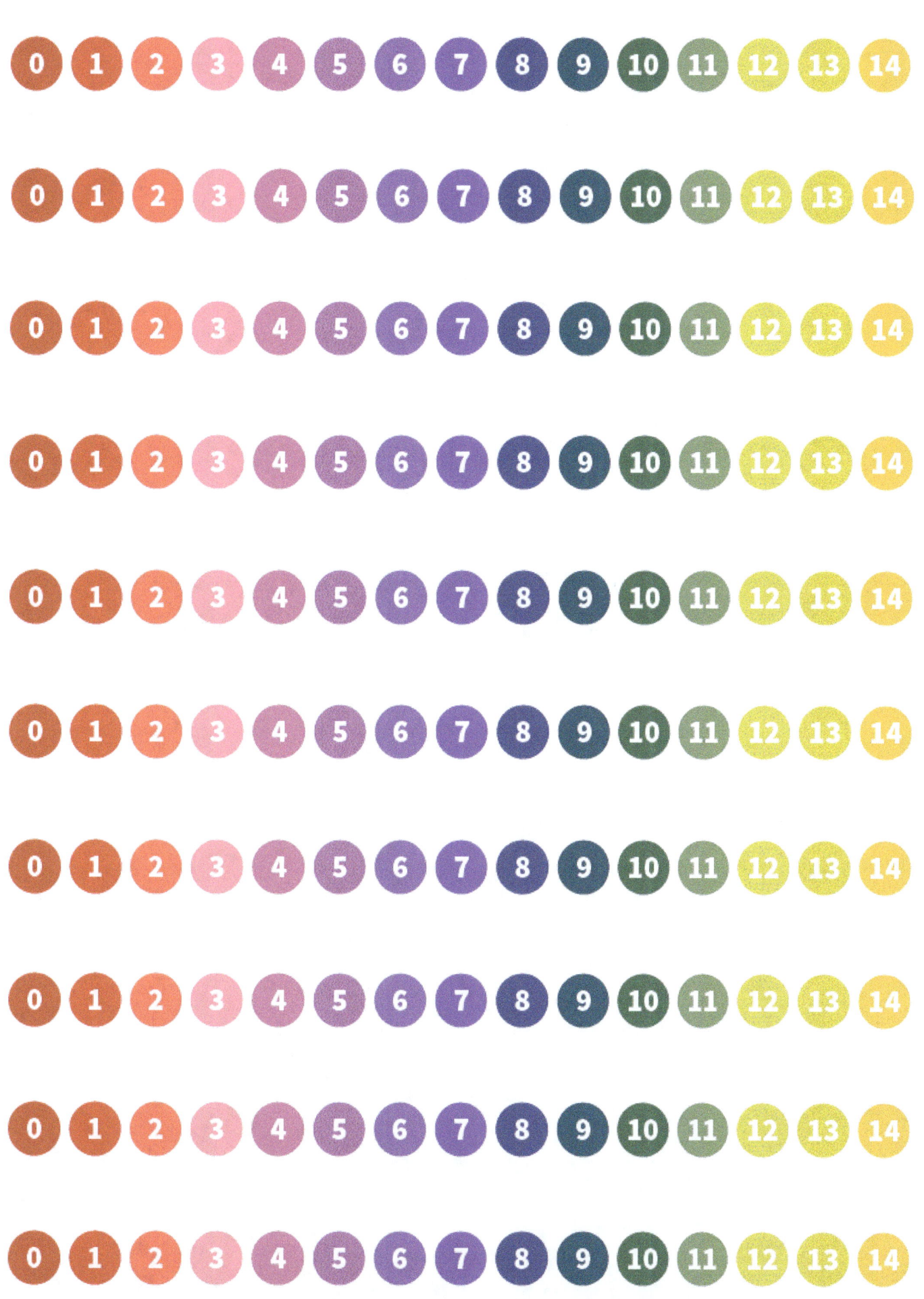

ACTIVITY IDEA 4F   EXPERIMENT: RIPENING FRUIT USING A NATURAL GAS

You will need:
- four green bananas (make sure that all four bananas are at the same stage of ripening)
- one ripe banana
- one piece of ripe fruit that isn't a banana
- four large plastic bags
- optional: a few small paper bags

What to tell the students:
In this experiment we will see how ethylene gas speeds up the ripening process in fruit. This gas is produced by fruit as one of the mechanisms that controls ripening. Commercial fruit growers can produce ethylene artificially and apply it to fruit that was picked unripe and therefore needs to ripen a bit before being sold to customers.

What to do:
1) One of your green bananas will be the "control" to show us how fast ripening occurs if we simply let the banana sit. Do not put this banana into a bag.
2) Put a green banana into a plastic bag and seal it shut.
3) Put the third green banana into a bag and add a ripe banana to the bag. Seal shut.
4) Put the fourth green banana into a bag with a piece of ripe fruit (apple, pear, etc.).
5) If you are using paper bags (to see if paper give a different result from plastic) repeat steps 2, 3 and 4 with paper bags.
6) Set all these bananas on a tray or in a box and put it in a place where it can sit undisturbed for a week.
7) Check the progress of the bananas each day. Record your findings so you don't forget!
8) As soon as all the green bananas have turned ripe, end the experiment.

Which banana was the slowest to ripen? Can you think of a reason? Did the plastic bags catch and hold ethylene and cause those fruits to ripen faster? Did the paper bags hold as much ethylene as the plastic bags?

"Ethylene" is one of the simplest hydrocarbon molecules.

$$\begin{array}{c} H H \\ \diagdown \diagup \\ C = C \\ \diagup \diagdown \\ H H \end{array}$$

ACTIVITY IDEA 4G   Test for antioxidant property of vitamin C

You will need:
- Lugol's solution (potassium iodide solution 3%)
- a piece of absorbent paper for each student
- paper towels
- vitamin C tablets
- a variety of foods to test (fruit or veggie slices, fruit juices, plus some things that might not have vit. C such as cheese, meat, milk-- but foods must be moist for the test to work)
- Optional: eye droppers for the test solutions
- Optional: a nitrile or latex glove for anyone applying the iodine solution to the paper (unless you can hold a crumpled paper towel in a way that doesn't get iodine on your fingers)
- protection for your work space (plastic covering or newspapers) so iodine solution does not stain the table (NOTE: We used a thin plastic tablecloth (the party kind) and it still leaked through and left yellow in the white plastic table.)

How to set up:
1) Being careful not to get the iodine solution all over your hands, use a cotton ball or crumpled paper towel to apply a small amount of Lugol's solution to the paper. (A little will go a long way, you don't need soak it.) Smear it all over the paper. The paper should turn black.
2) Crush the vitamin C tablets and put the powder into a little water to make a concentrated solution.
3) Have all your fruits and other food samples cut into flat, thin slices, or, if liquid, into small cups (provide eye droppers)

What to tell the students:
   In this experiment we will see how ascorbic acid (vitamin C) acts as an anti-oxidant and reverses the process by which iodine stains things black. The ascorbic acid, $C_6H_8O_6$, will donate a hydrogen atom to each iodine atom. When the hydrogen atoms bond to the iodine atoms, it will change how the iodine reacts to light, and therefore produce a color change. We will put samples of various foods onto the iodine-soaked paper to see which ones contain enough vitamin C to produce a color change.

What to do:
1) First, do a control test using water. Does water change the color?
2) Now try the concentrated vitamin C solution. Put a few drops onto the paper. You should see a dramatic change from black to almost white, or the original color of the paper.
3) Now lay samples of various foods onto the paper. Press them down if they are not in contact with the paper. Do any of them create white spots on the paper? Which foods surprised you one way or the other?

The chemical formula for this reaction was given as the following (from the noted website below):
($C_6H_8O_6 + I_2 \rightarrow C_6H_6O_6 + 2HI$

If you want more information or to see a very short video of this lab, go to:
https://melscience.com/AT-en/articles/how-test-vitamin-c-home/

ACTIVITY IDEA 4H   SNACK: TRY SOME TEAS

Tea is the most consumed beverage in the world (not counting water). All types of tea (green, black, white) are from the same tea plant, *Camellia sinensis*. It is the processing that gives various types of tea. Green tea is less processed than black tea. **Check out the tea growing videos on the youtube playlist**. You will find that there's more than one way to process tea. They do it differently in each video.

If this idea works in your situation, you might even want to use fancy tea cups and saucers and do a little education about etiquette, or about tea customs in various countries.

If you want to try decaffeinated teas, there is also a short video explaining the ways of decaffeinating tea and coffee.

If you want to add a food to eat with the tea, consider providing some berries that are high in the phytochemicals we read about—flavonoids, anthocyanin and resveratrol.

ACTIVITY IDEA 4I SNACK: TRY SOME SALAD GREENS

There are many types of salad greens, including many varieties of lettuce, plus leaves from spinach, kale, and beets, as well as collard and mustard greens. Your geographical area may have unique greens not available in other places. Provide a selection of samples for students to try. You might also want to provide a small amount of a few varieties of dressings for dipping the leaves. After all the taste tests, take a vote to see which green is the favorite and which is the least favorite.

ACTIVITY IDEA 4J   EXPERIMENT: REGROW A HEAD OF LEAF LETTUCE

You will need:
- a head of lettuce that is long and straight (such as Romaine)
- a saucer or small bowl
- fresh water each day

What to do:
1) After you've cut off the leaves, don't throw away the stump! Put it into a saucer or bowl and add enough water to cover the bottom of the stump.
2) Change the water each day, and see what happens over the course of a week.
3) Hopefully, new leaves will grow and you'll be able to harvest them for your salad or sandwich. Look at the base of the lettuce. You will see tiny roots growing. These roots aren't quite the same as the original roots, but they take in enough water for a few new leaves to grow.

ACTIVITY IDEA 4K MOLECULE MAT for chapter 4

You will need:
- a copy of the following pattern page for each student
- toothpicks
- the materials for the atoms

What to do:
1) Put your chosen materials inside the boxes on the left side of the page (or in small dishes if they won't fit inside the boxes). Toothpicks can be set in a dish, or simply in a pile, within the student's reach.
2) Let the students work on their own as much as possible.
3) The molecules are likely to be large enough that all four will not fit onto the page. You can tell your student to build two then recycle the items, or give them an extra blank sheet and have them build several molecules on it.
4) For students who are keeping a portfolio of their work, take a photo of their paper with all the finished molecules on it.

MOLECULE MAT Chapter 4

resveratrol

pyridoxine, vitamin B6

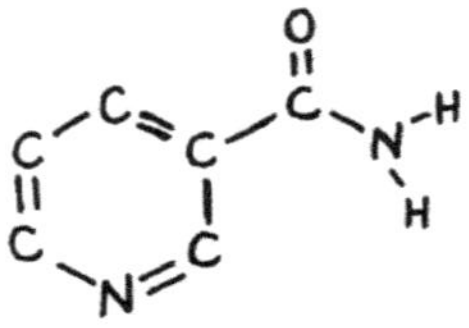

nicotaminide, a form of vitamin B3

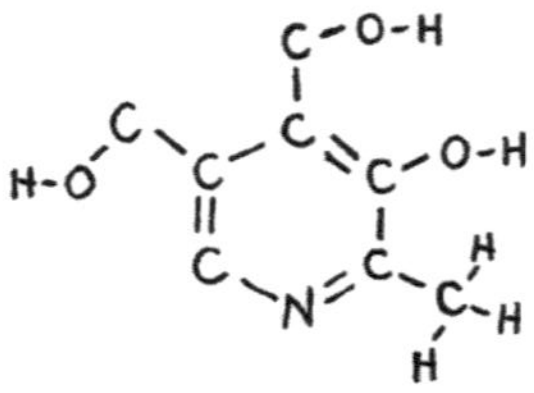

phenol

H - hydrogen	
O - oxygen	
C - carbon	
N - nitrogen	

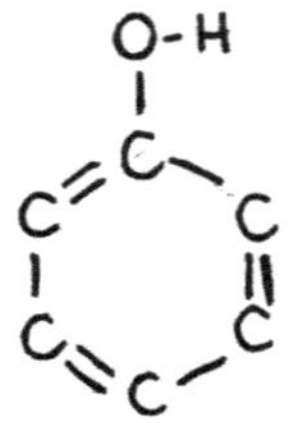

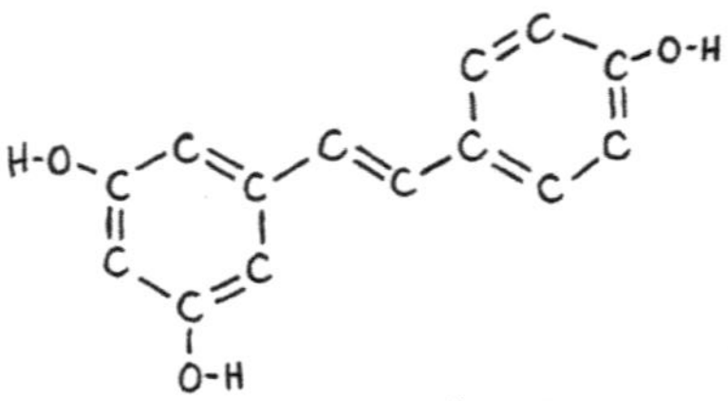

ACTIVITY IDEA 4L   REVIEW BINGO for chapter 4

You will need:
- a copy of the following pattern page for each student, and scissors to cut them apart

What to do:
1) Make a copy of the following page of squares for each player. Cut them apart.
2) Players will then choose 16 of their cards to form a 4x4 grid. Make sure to use vitamins A, B, C and D.
3) Caller will choose clues to read. Please note that some clues have more than one answer. Also note that there are several clues for each word, so you can have fresh clues for a new game. But if you have to use a clue twice, no problem, review is great.
4) When someone gets Bingo, start over and rearrange your grid, but keep vitamins A, B, C, and D in the grid.

CLUES
1) Our skin makes this vitamin in response to exposure to UV light. (D)
2) A vitamin found in cod liver oil (A or D)
3) A fat-soluble vitamin (A, D or K)
4) This vitamin is needed in the blood clotting process (K)
5) A lock of this vitamin causes rickets (D)
6) The mineral element that gives vitamin B12 its reddish pink color. (cobalt)
7) A string of glucose molecules with their flags all pointing in the same direction (starch)
8) A string of glucose molecules where every other flag is pointing in the opposite direction. (cellulose or fiber)
9) A string of glucose molecules that we cannot digest. (cellulose or fiber)
10) A string of glucose molecules that we can easily digest (starch)
11) Another word for cellulose when it is found in our diet. (fiber)
12) A molecule with three fatty acids hanging from a glycerol hanger. (triglyceride)
13) The smallest amino acid   (glycine)
15) An amino acid that can make bridges between proteins strands using sulfur bonds   (cysteine)
14) This vitamin can also be called ascorbic acid (C)
15) A lack of this vitamin can cause an illness once known as pernicious anemia (B12)
16) A lack of this vitamin causes scurvy (C)
17) The element that sits in the middle of the chlorophyll molecules (magnesium)
18) This vitamin can be made from beta-carotene (A)
19) This molecule is made of a chain of carbon atoms what have hydrogens attached to them. (Fatty acid)
20) A molecule in a leaf that can catch energy from sunlight (chlorophyll or beta-carotene)
21) Plants use these molecules to kill or deter the bigs that eat them. (phytonutrients)
22) This molecule has a carboxyl group on one side and an amine group on the other. (cysteine, glycine)
23) This molecule is like a little task robot that does one particular job (enzyme)
24) This molecule fits into a tiny gap in an enzyme molecule (coenzyme)
25) Many vitamins function as this type of helper molecule. (coenzyme)
25) These molecules protect us against harmful broken molecules called free radicals (anti-oxidants)
26) This molecule is what all membranes are made of (phospholipid)
27) This molecule has a polar head and two non-polar, fatty tails (phospholipid)
28) This type of molecule helps polar and non-polar substances mix together in a liquid (emulsifier)
29) In mayonnaise, both eggs and mustard function as (emulsifiers)
30) This molecule turns red in acidic environments and blue in alkaline environments. (anthocyanin)
31) This molecule gives both red and blue fruits their colors. (anthocyanin)
32) Vitamin A and vitamin K molecules have a long, skinny section that looks like a (fatty acid)
33) We cannot digest this type of molecule, but the bacteria in our gut love to eat it. (fiber or cellulose)
34) A water-soluble vitamin (B12 or C)
35) Vitamin B12 is also called cobalamin. This name comes from the element (cobalt).
36) This molecule resembles a barbell. (beta-carotene)

phospholipid	coenzyme	vitamin D	vitamin A
vitamin C	vitamin B12	anti-oxidants	phytonutrients
fatty acid	vitamin K	emulsifier	beta-carotene
chlorophyll	triglyeride	cellulose	starch
cellulase	cysteine	cobalt	fiber
magnesium	anthocyanin	glycine	enzyme

CHAPTER 4 Bingo cards   Print one copy per player.

CHAPTER 5

ACTIVITY IDEA 5A   LAB: DISSECT SOME BEANS

You will need:
- a selection of beans or peas (green beans, snap peas, even some dried beans that have been soaked)
- optional: You can provide utensils for dissecting beans, but fingers can do a pretty good job, too.

What to tell the students:
   This is your chance to see those parts you read about in the chapter. Can you find the plumules and embryo in each seed? Can you peel off the seed coat? Can you find the cotyledons?

What to do:
1) If working with pods, peel the pods open slowly and carefully so you can see how the seeds are attached. (After the beans detach from the pod, the scar left on the bean (where it was once attached) is called the hilum.)
2) Are all the seeds the same size? Sometimes you'll find a "runt" that seem to not be developing much at all compared to the other seeds in the pod.
3) Open each bean or pea seed carefully and find the plumule leaves, the two cotyledons, and the radicle (the future root).

ACTIVITY IDEA 5C   SNACK: TRY SOME PLANT-BASED "MEATS"

   Buy a selection of different plant-based meats and cook according to package directions. Have the students read the ingredients labels to get an idea of what is in each product. Have the students discuss their reactions to each food. Is the company trying to fool you into thinking you are eating meat, or have they made a product that stands on its own as a unique food product?

ACTIVITY IDEA 5C   SNACK: TRY SOME BEANS

   Buy one can each of as many types of canned beans as you can find (dark kidney, light kidney, black beans, cannellini, fava or butter beans, navy beans, mung beans, black-eyed peas, great northern beans, chickpeas (a.k.a. garbanzo beans), pintos, limas). Drain the beans (keep the liquid if you want to do activity 5D, making "aquafaba"). Have the students compare taste and textures. Which one holds its shape best? Which one has the strongest flavor? The least flavor? Have them vote for their favorite, or least favorite, bean.

ACTIVITY IDEA 5D   COOKING ACTIVITY: MAKE "AQUAFABA" (an egg-free alternative meringue)

You will need:
- the liquid from a can of chickpeas (also called garbanzo beans) (If you can't use chickpeas, substitute with liquid from another type of bean, such as kidney beans)
- cream of tartar
- granulated sugar
- vanilla (either real or imitation)
- an electric hand mixer or a stick blender

What to do:
1) Put the bean liquid into a bowl. Add 1/4 teaspoon (1 g) cream of tartar. (If using the liquid from more than one can, use 1/4 teaspoon for each can.)
2) Beat this mixture until it is foamy and soft peaks start to form.
3) Add the sugar gradually, while beating.
4) Add vanilla, and beat until stiff (pointed) peaks form.
5) Now it is ready to use. If you don't have a pie to put it on, you can bake it into cookies. Put spoonfuls of meringue onto a cookie sheet and baked on greased cookie sheet at 200° F (95° F). Bake for 90 minutes, until crisp and dry.

ACTIVITY IDEA 5E   PLAY MORE "FRUIT AND VEGGIE CARD GAMES"

Hopefully, you didn't play all the games already! Since we studied beans and potatoes in this chapter, vegetable cards fit right in with the theme of this chapter. Play at least one more game with the cards.

ACTIVITY IDEA 5F   REVIEW CARD GAME: "TEAM UP FOR TEN"

You will need:
- copies of the following pattern pages (printed onto card stock if possible)
- scissors
- tape or string to divide your playing area

How to prepare:
1) This game is played with two teams. A team can be from one to four players (though in a pinch you could have as many as six on a team.) You don't need the same number of payers on each team. For example, you can have three on one team and four on the other. You will need two sets of cards (two copies of each of the three pattern pages) if you have a total of two to six students playing. Make three copies of each page if you have a total of seven to twelve players.
3) Use string or tape to divide your playing space. (I used blue painter's tape on my long plastic table.) You need a three areas, as shown. The middle strip is a neutral zone that does not belong to either team. All the cards will be scattered, face down, around the length of the neutral zone at the beginning of the game.

How to prepare:
1) Before you start to play, let the students see all the cards. Help them to find all the matches.
Here are the molecules you can make:

1) amino acid (the only set that has three cards instead of two)   2) acetic acid (C_3COOH)
3) sodium bicarbonate $NaHCO_3$ 4) calcium phosphate $Ca^{2+}PO_4^{3-}$
5) triglyceride (fatty acids plus glycerin "hanger")   6) water, H_2O
7) potassium benzoate   8) beta-carotene (one bunny card will be upside down)
9) sucrose (gray carbon atoms)   10) lactose (white carbon atoms)

2) Pull out these cards: lactase enzyme, sucrase enzyme, beat-carotene enzyme, pH card. These four cards are able to split molecules. Make sure the players can identify which molecules they can split (lactose, sucrose, beta-carotene, water).
3) Stack all the cards together and shuffle thoroughly.

How to play:
1) The object of the game is for your team to work cooperatively to form all ten molecules listed above. You will lay your finished molecules out on the table so both teams can see them. The first team to make all ten molecules wins the game.
2) Shuffle the cards well, then scatter them, face down, along the neutral zone.
3) Each player then draws 2 cards to start. Each player's hand of cards can seen by other members of their own team, but should be kept hidden from the other team.
4) The teams will take turns playing. During a team's turn, each player on that team will choose ONE of the following actions.
   OPTION #1: Draw a card from the pile in the neutral zone.
   OPTION #2: Trade one card with another player on your team. When this option is used, it will count as both players' chosen option. This option is used in order to bring pairs of cards together so you can lay them down. As you play, you will discover that you often need to look at everyone's hands to determine what the best trade will be. Each player can only trade one card per turn, so you can't have three players all trading cards. If two players decide to trade cards (in order to put a match in one person's hand), then those two players have used their move for that turn and a third player will have to make a move on their own. As the game progresses, you will begin to see trading strategies that are more advantageous than others.
   OPTION #3: Play a TRADE card if you have one. The TRADE card allows you to return one of your cards to the neutral zone, along with TRADE card, then draw a fresh card.
   OPTION #4: Play an enzyme card or a pH card against the other team. For example, if the other team has a completed a sucrose molecule, and you have a sucrase enzyme, you can destroy their molecule and return the parts to the neutral zone. Once a destruction card (enzymes and pH) is played, it is removed from the game.

5) The only "free" move that does not count as your chosen option is laying down a completed set/pair of cards. For example, you can choose the option to draw a card from the neutral zone, and if it happens to match a card in your hand, you can immediately lay down the finished molecule.

Adaptation of the rules for playing with only two players (one against one):
1) Start with four cards instead of two.
2) Obviously, you can't share or trade cards with another player so you can't use option #2. However, the game will still work even without this option.

ACTIVITY IDEA 5G   Testing taste buds

You will need:
- cotton swabs
- 5 small cups of water
- salt
- sugar
- baking soda (for bitter taste)
- citric acid (or substitute lemon juice)
- MSG (often sold as "Accent" in spice section)

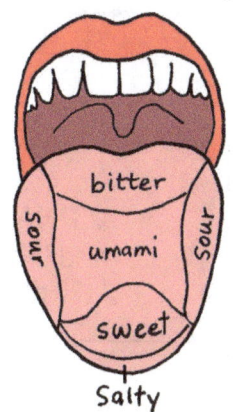

Maps like this are all over the Internet. You can easily find one to download and print for your students.

What to tell the students:
　　　In this experiment we will test the idea that our tongues have particular areas for particular tastes. There are only 5 tastes that our tongues can sense: salty, sweet, sour, bitter, and savory (umami). The last one was officially discovered by a Japanese scientist in the early 1900s. He was able to isolate the molecule responsible for this taste sensation: monosodium glutamate, or MSG. Monosodium glutamate is the amino acid glutamate with a sodium atom attached to it. The sodium atom comes off as soon as the MSG is dissolved into our saliva. Glutamate is an amino acid that our bodies use as a communication molecule. The sensory cells in our taste buds release glutamate in order to send a signal to a nearby nerve cell. The nerve cells then carry the signal to the brain. So if we eat foods high in glutamate, such as cheese, mushrooms, or tomatoes, our brain gets a strong signal from those savory sensors in our taste buds.
　　　Because MSG tingles the savory receptors in our taste buds, it can be added to bland foods in order to make us think they taste really good. Junk foods such as Dorito® chips are coated with MSG. That's why it is hard to stop eating them!
　　　You will test various spots on your tongue to see which areas are most sensitive to sweet, sour, bitter, salty and savory.

What to do:
1) Use these materials to make small cups of solutions: salty, sweet, bitter, sour, savory (umami) The exact amount of solute (sugar, salt, etc.) is not important. Make sure the mixture is strong enough. Make sure you have the cups labeled.
2) Look at a tongue map, like the one shown here. (Tongue maps are easily found on the Internet.) Notice which areas are thought to have more sensors for the various tastes.
3) Dip a cotton swab into one of the solutions and then put it on various parts of your tongue. Do your results agree with this map? There isn't a right answer here. Your tongue might be slightly different than someone else's.
NOTE: Make sure not to dip your used swab into another cup! Use a fresh swab for each dip.

ACTIVITY IDEA 5H   MOLECULE MAT for chapter 5

You will need:
- a copy of the Molecule Mat chapter 5 pattern page
- toothpicks
- the materials for the atoms

What to do:
1) Put your chosen materials inside the boxes on the bottom of the page (or in small dishes if they won't fit inside the boxes). Toothpicks can be set in a dish, or simply in a pile, within the students' reach.
2) Let the students work on their own as much as possible. The molecules are likely to be large enough that all both will not fit onto the page. You can tell your student to build one, then recycle the parts. Notice that we are not going to worry about making these molecules curve. We'll just focus on counting the C's and H's.
4) For students who are keeping a portfolio of their work, take a photo of their paper with all the finished molecules on it.

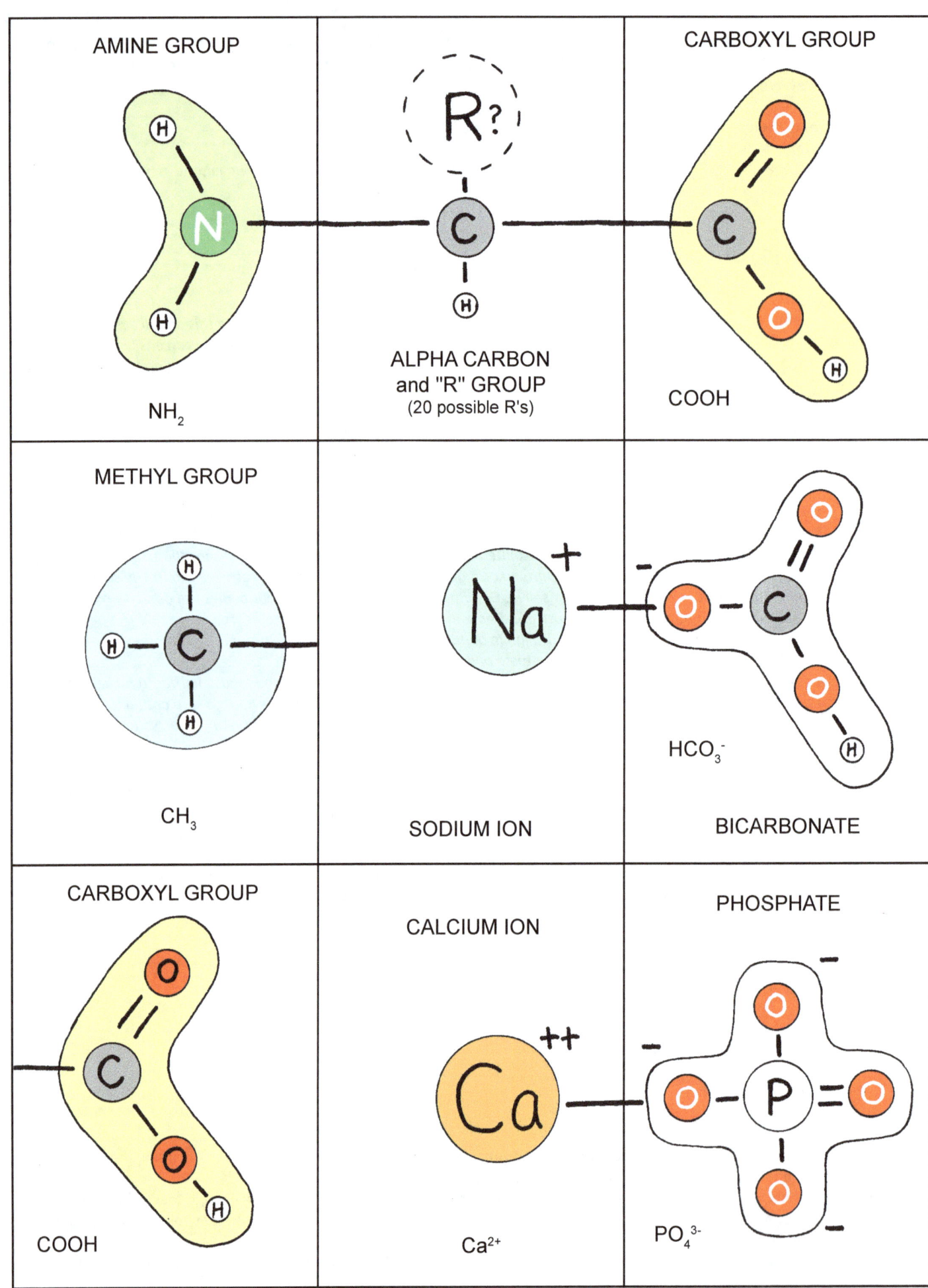

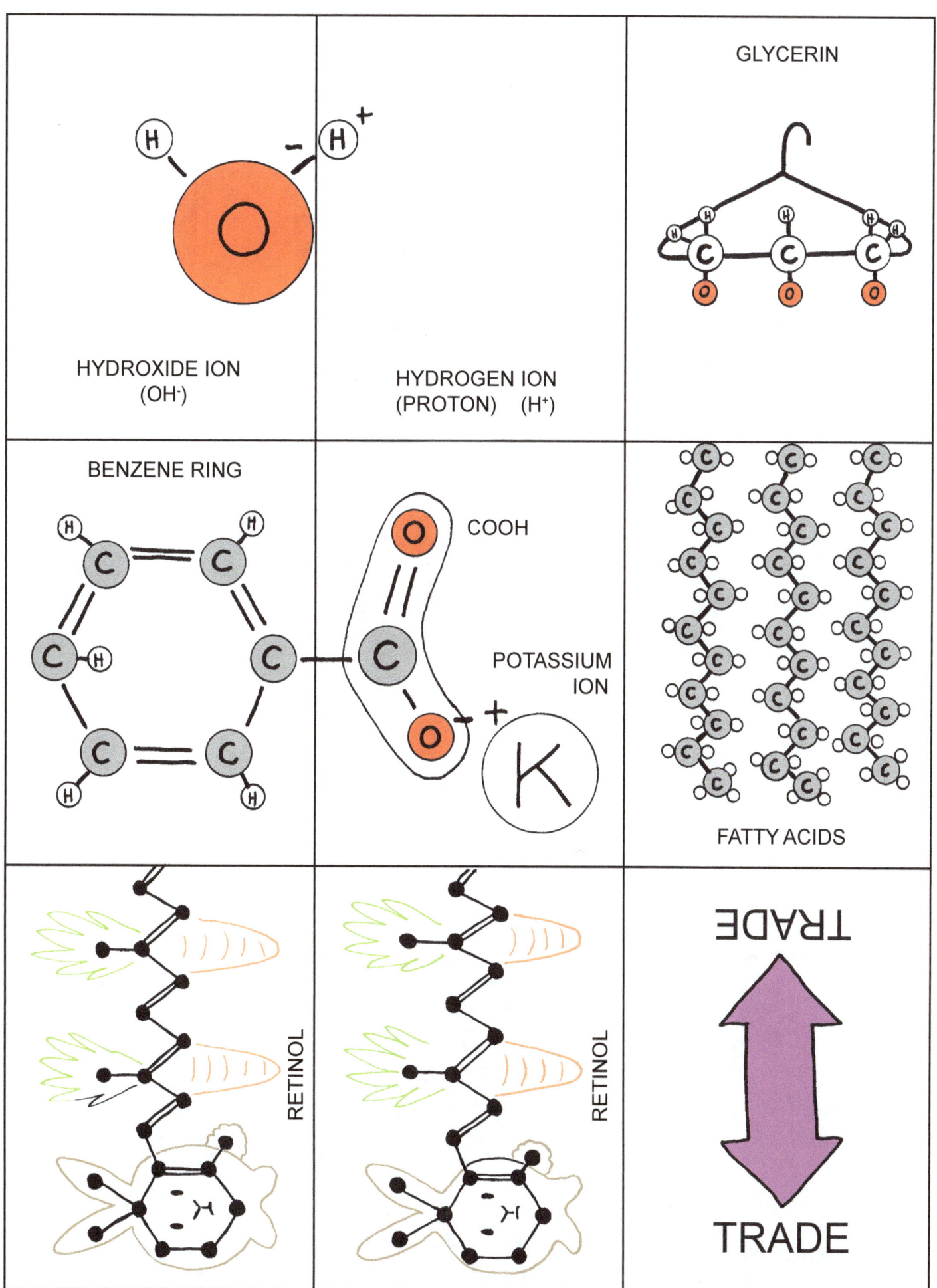

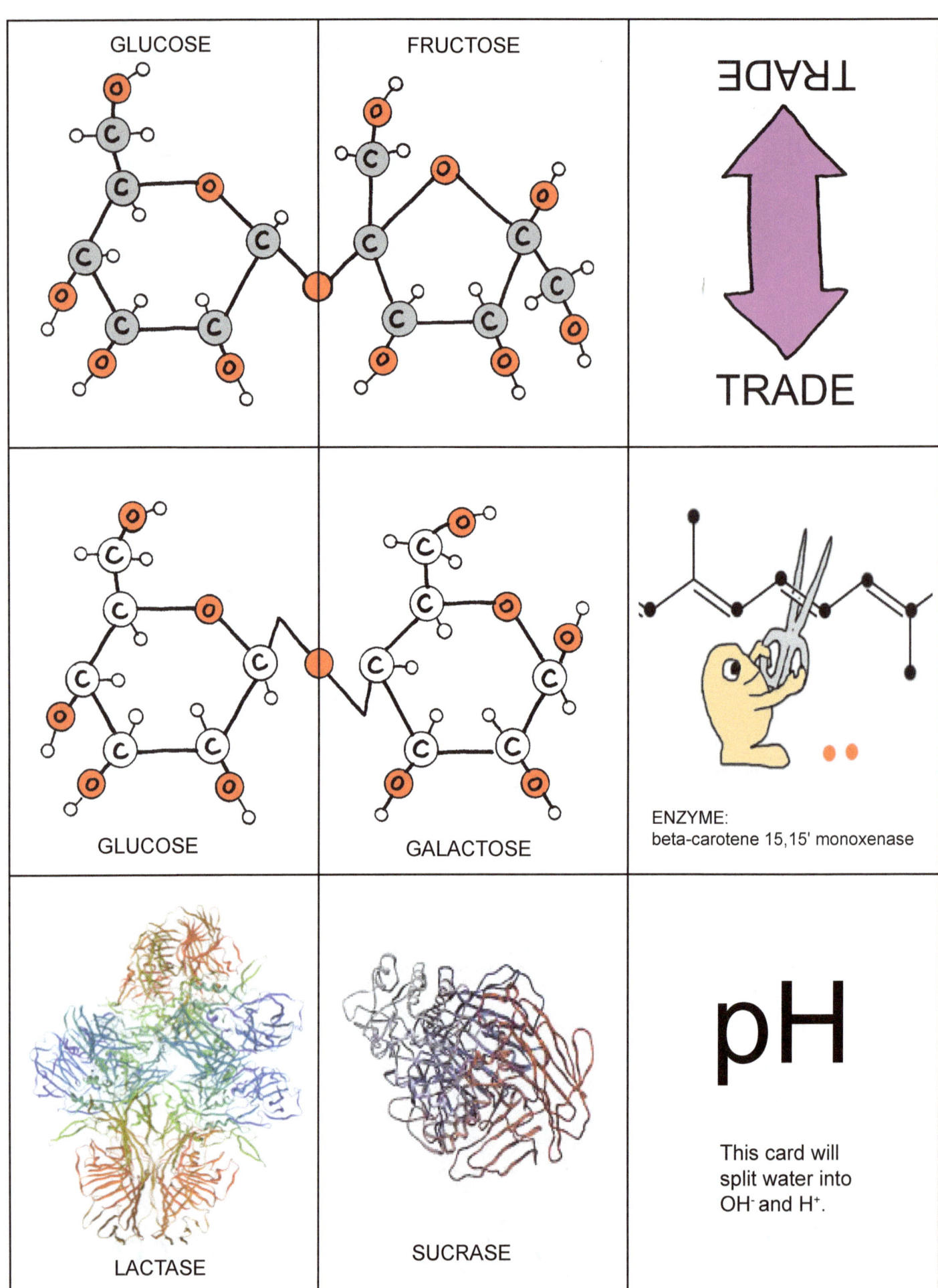

MOLECULE MAT   Chapter 5

Omega-6 fatty acid
LINOLEIC ACID *(lin-oh-LAY-ick)*

EPA Omega-3 fatty acid
(Eicosapentanoic acid)

COOH is marked just to help you see it as a unit. That H might come off, which is why it's called an acid.

We saw in the chapter that this molecule is actually curved. However, it will be easier for you if we don't worry about the curve and let you just concentrate on where the H's are.

| H - hydrogen | O - oxygen | C - carbon |

CHAPTER 6

ACTIVITY IDEA 6A   EXPERIMENT: HOW MANY DROPS OF OIL WILL FIT ON A PENNY?

You will need:
- an eye dropper for each student
- a penny for each student
- vegetable oil (perhaps a few different types of you have them on hand)
- a cup of water (as it is probably a good idea to revisit the first lab where we did water on a penny)

What to tell the students:
　　Way back in chapter one, we did an experiment where you found out how many drops of water would fit on top of a penny. We noted that water is a polar molecule and therefore creates hydrogen bonds between the molecules. These bonds are relatively weak, but are strong enough to hold the molecules together so that a fairly large drop of water can be created on top of a penny. In this chapter we discussed different kinds of vegetable oils, especially soybean oil. Oils are non-polar molecules and therefore don't experience hydrogen bonding. Will their lack of hydrogen bonding prevent them from being able to form a large droplet on a penny?

What to do:
1) Give each student a penny, an eye dropper and access to a cup of water.
2) Have them do the water drop challenge again. They will likely get 30 or more drops to fit.
3) Dry off the penny and gives them access to the vegetable oil. Have them count drops of oil till the oil slides off the penny. They will probably get less than 20, perhaps even less than 15.

EXTENSION: Ask them to draw a rough picture of what the oil molecules look like. (They should draw long lines, perhaps zigzag lines, or even curved lines. They might include the letters C and H. Drawing triglycerides with three legs would also be acceptable.)

ACTIVITY IDEA 6B   LAB: DISSECT AN EGG   (NOTE: Activity 6C will suggest one way to use your cracked eggs.)

You will need:
- an egg
- a small plate
- optional: craft sticks or dull table knives (for poking and prodding egg)
- optional: magnifying glass for looking at shell (highest power possible)
- optional: craft knife with very sharp point (for supervising adult, or for older students)

What to tell the students:
　　We crack eggs and use them in baking all the time. But have you ever stopped to really look at them? In this activity you will do a scientific examination of an egg and try to find the parts shown in the diagram on the first page of this chapter.

How to prepare:
1) Buy extra large or jumbo eggs. If you can get some farm-fresh eggs, that's even better.

What to do:
1) If you have a magnifier, look at the shell. Is the texture very smooth or does it look rougher at higher magnification? (The actual pores in the shell are microscopic.) If your egg has never been washed, it will still have its natural cuticle coating that helps to keep bacteria out. Fresh eggs do not need to be refrigerated. If your egg is from a store, it has probably been washed so the cuticle layer is gone. Washed eggs need to be kept cold.
1.5) OPTION: You may want to tap a tiny hole in the large end of the shell. This requires a small pointed object, such as the tip of an X-Acto knife. If working with younger students, the supervising parent or teacher might want to do this operation on each egg. If done properly, the hole should open into the air space and you will be able to hold the egg with the hole pointed down and nothing will leak out. You can see why this works of you look at the picture of the egg anatomy. The air sac is sealed off from the contents of the egg. (Only the large end will have this air space.)
2) GENTLY crack the egg and allow the insides to slip out onto a plate. Try not to injure any of the inside parts as you do this.

3) Examine the shell. Look at the large end and find the air pocket. As an egg ages, this air space grows larger. Examine the thin membrane that covers the inside of the shell. Peel off a bit of the membrane and give it a tug. It will be surprisingly strong. There are actually two membranes there, but we see them as one.
4) Observe the egg white, also called the albumen. Can you see any variation in the clear gel? Very gently poke various places around the white. Are there places that seem more firm? Places that seem more runny? The albumen actually has several distinct sections, but because all these sections are clear, they are hard to differentiate just by looking.
5) The chalaza will be easy to find. Observe both sides. Is one longer than the other? If you tug on it gently, can you see that it is attached to the yolk?
6) Observe the yolk carefully without breaking it. Can you find a spot? In some eggs there might be a fleck of red or orange on the yolk, but in other eggs it could be just a lighter spot. This is the germinal disc, where a chick would grow IF the egg was fertilized by a rooster and IF the egg is kept at the right temperature.
7) The yolk is surrounded by a thin membrane called the vitelline membrane. Vitelline means "glass-like" (clear). Gently poke the vitelline membrane in one place. Can you see a hole open up? The liquid yolk will come out through that hole.
8) The eggs can then be cooked and eaten as a snack, or you can use them in the recipe below.

EXTENSION: Find some other types of eggs to compare to a chicken egg (duck, quail, goose, etc.) If you don't live in a rural area, try an Asian grocery store, as they often sell specialty eggs.

ACTIVITY IDEA 6C   COOKING ACTIVITY: MAKE "FLAN" (also known as "crème caramel")

This dessert originated in Europe but is now served all over the world. The word "flan" comes from an old French word for "flat cake." Each country has its own variation; sometimes only yolks are used, sometimes sweetened condensed milk is used, some prefer coconut flavoring instead of vanilla, some like their flans to have a thin texture, others prefer thick. (For a complete list of regional variations, see the Wikipedia article on "Crème Caramel.")

You can use the recipe listed here, or you can surf the Internet, or a video steaming service, to find one more to your liking. (NOTE for those of you who don't consume sugar: You can't make caramel using a sugar substitute, but if you need a sugar-free option, you can skip making the caramel yourself and look for a sugar-free caramel sauce at your grocery store. Pour this on the bottom of your pan. You can use a sugar substitute in the egg and milk mixture.)

This activity reinforces the information about eggs (pages 100-101), and the information about caramel (pages 110-111).

You will need:
- a medium-sized shallow baking dish
- a large, shallow pan that the baking dish will fit into
- a non-stick, or stainless steel, frying pan (medium size) for making the caramel sauce
- a serving platter larger than the bottom of your baking dish
- optional: a pastry brush and a bowl of water
- 3/4 cup (150 grams) granulated sugar
- 4 eggs (you can use the eggs you dissected)
- 1/2 cup (100 grams) granulated sugar
- pinch salt
- 2 teaspoons (10 ml) imitation vanilla, or 1 teaspoon (5 ml) real vanilla extract (real vanilla tends to be stronger)
- 2 cups (500 ml) milk (or a combination of cream and milk)

What to do:
NOTE: If you'd like to simply follow a video instead of these written instructions, video streaming services (e.g. YouTube) have many posts that show people making flan.
1) Pre-heat oven to 350º F (175º C).
2) Pour 3/4 cup (150 grams) granulated sugar into the frying pan. Some people don't add any water, they just cook the sugar as is. This can be an interesting science phenomenon to watch, as the dry sugar crystals melt into a liquid. However, you can also add 1/4 cup (60 ml) water. Pour the water slowly, distributing it as evenly as possible around the sugar. You can tip the pan and try to swirl the mixture a bit, but don't stir.
3) Set the pan on a burner with medium heat. You can shake the pan back and forth, or swirl it around, as the sugar cooks, but you are not supposed to stir. You can dip the pastry brush into water and brush around the inside edge of the pan, just above the cooking sugar. However, I watched some videos where they did not use a brush and their results were fine, too.

4) The sugar will become more liquid and begin to boil. Swirl the mixture but don't stir and let it cook like this until it turns light brown. Remove from heat and pour the sauce into the bottom of your baking dish. (You can use a scraper to get all of the caramel out of the pan.)
5) Put the eggs, 1/2 cup (100 grams) sugar, a pinch of salt, and the vanilla into a blender, or into a bowl if you are using a hand mixer. Beat for half a minute or so.
6) Add 2 cups (500 ml) of milk (or a mixture of milk and cream) and beat for another half minute or so.
7) Pour this mixture into the baking dish, on top of thin layer of caramel sauce.
8) Set the baking dish inside a larger pan and add water to the large pan until half the height of the baking dish is covered.
9) Place in oven and allow to bake for about 45-50 minutes. (Remember, at the top of page 101 we discussed how egg yolks slow down the coagulation process and therefore egg-thickened desserts must bake for a longer time than other recipes.) Begin to check it at the 40 minute mark, and every 5 minutes or so after that. The flan is done when you can insert a knife and it looks clean when you pull it out. If it comes out with white or yellow on it, let the flan bake a little longer.
10) Allow the flan to cool for a short time. Take a knife and slide it around the top edge of the flan to loosen the edges from the baking dish. Then put the serving platter on top of the baking dish and carefully turn them over so the baking dish is upside down on the platter. Carefully lift the baking dish. The flan should now be on the platter with its caramel sauce on the top. Allow to cool before serving. The flan can be stored in the refrigerator, but don't cover with plastic wrap, as the wrap will stick to the caramel and pull it off when you remove the wrap.

ACTIVITY IDEA 6D   EDIBLE EXPERIMENT: AN ENZYME THAT PREVENTS GELATINIZATION

The fact that fresh pineapple contains an enzyme (bromelain) that prevents jello from thickening is well-known and likely you or your students may have experimented with this already. I will suggest a few extra unknown factors that will enable you to use the idea even with students who already know what bromelain does to jello.

You will need:
- collagen-based gelatin, either sweetened and flavored (ex: Jell-O®) or plain (ex: Knox® unflavored gelatin packets)
- agar (also called "agar agar") which can be ordered online from places like Amazon, or purchased from Asian groceries
- optional: a Kosher (fish-based) gelatin, or a vegan gelatin mix
- a fresh pineapple
- another type of fresh fruit (NOTE: Papaya and kiwi will give similar results to fresh pineapple, so you might want to choose something like bananas, apples or oranges. However, using papaya and kiwi could add another layer of complexity for students who have already done pineapple.)
- meat tenderizer
- hot water
- small containers, or a few ice cube trays (Ice cube trays work well when you have many students who will all want to sample the final results.)

What to tell the students:
In this activity, you will experiment with an enzyme that breaks down proteins, including the collagen found in jello. You will find out if heat has any effect on the enzyme, and you will test whether this enzyme affects non-collagen gelatins.

What to do:
1) Chop the pineapple into bite-sized pieces.
2) Put a few dozen pieces into boiling water and let them cook for 5 minutes. Drain and cool.
3) Mix your gelatin(s) according to package directions. (The directions on my bag of agar suggest using 1 teaspoon (5 ml) of powdered agar for each cup (250 ml) of water. Allow the agar powder to soften and dissolve completely.)
4) Label your containers or cubes with the following categories. (The number of cubes slots for each category is up to you.)

    1) plain gelatin (your "control")
    2) gelatin with fresh pineapple
    3) gelatin with cooked pineapple
    4) gelatin with another type of fruit
    5) plain agar
    6) agar with fresh pineapple
    7) agar with cooked pineapple
    8) agar with other fruit   10-12) optional-- another type of gelatin plain,
    9) gelatin with meat tenderizer added   with fresh pineapple, with cooked pineapple

5) Pour the correct solutions into each container and refrigerate all of them until the plain gelatin in #1 is firm.
6) Pull all samples out of the refrigerator and observe them. Which ones did not gel?

EXPECTED RESULTS: The collagen-based gelatin (jello) with the fresh pineapple will not gel. This is because pineapple contains an enzyme called bromelain that breaks down proteins including collagen. Bromelain cuts the long collagen strings into short pieces. The gelling process requires the collagen to form long strings in order to hold the water. Bromelain prevents this from happening. Like many enzymes, bromelain is affected by heat. The heat denatures the enzyme and makes it unable to do its job. Therefore, the cooked pineapple sample should gel. Meat tenderizer contains bromelain, so that sample should yield the same results as the fresh pineapple. Agar is not made of collagen, so we can expect it to gel even with the fresh pineapple. Kiwi and papaya also contain protein-digesting enzymes so if you used these fruits, they will probably not gel, or will form only a weak, watery gel.

ACTIVITY IDEA 6E   COOKING ACTIVITY: MAKE MERINGUE

Find a recipe in a cookbook or online, and experience the science of meringue for yourself. You can put your meringue on top of a pie (perhaps a lemon meringue?) or just enjoy it as a fluffy treat.

ACTIVITY IDEA 6F   SNACK: MAKE ICE CREAM IN A PLASTIC BAG

There are many videos online showing how to do this activity. Basically, you put two cups of a milk liquid of your choice (milk, cream, half & half, coconut milk, almond milk, etc.), some sweetener of your choice, and some flavoring of your choice into a gallon size "ziplock" bag. Fill another gallon size ziplock bag with ice cubes (2/3 full) and add half a cup of salt (rock salt works best but table salt will be okay, too). Make sure the milk bag is zipped tightly shut, then place it in the middle of the ice cubes, then zip the ice bag shut. Put a towel around the ice bag so your hands don't get too cold, and shake the bag vigorously for about 5 minutes. When you open the milk bag you should find frozen ice cream.

ACTIVITY IDEA 6G   COOKING ACTIVITY: MAKE "POPPING BOBA" (gelatinized spheres, the kind in bubble tea)

This activity requires some special chemicals that you will have to order. Amazon carries them, but Asian groceries or science supply companies may also, as well. These ingredients are a little expensive, so you will have to decide whether it is worth it for you. If you want a less expensive lab that is very similar and uses things you probably have on your shelf, see activity 6F. (NOTE: Some bubble teas use tapioca spheres instead of sodium alginate. This type of sphere is chewy and does not pop.)

Instead of trying to reprint all the instructions I've read on various sites, it is probably better for you to read them yourself. Try one of these addresses, or use key words "popping boba" in an Internet search engine, or search YouTube or another video streaming service.

https://www.honestfoodtalks.com/popping-boba-diy-recipe/   (LOTS of information, pictures, and embedded video clips.)
https://kitchenpantryscientist.com/tag/popping-boba/   (This site has less info, but suggests also making larger bubble that they call edible water balloons.)

You will need:
- calcium chloride or calcium lactate (special order it)
- sodium alginate (special order it) This is a natural chemical derived from seaweed.
- distilled water
- a large dropper or syringe (can use a baster)
- a slotted spoon
- fruit juice

ACTIVITY IDEA 6H LAB: ALTERNATIVE RECIPE FOR GELATIN SPHERES (technically edible, but not designed for eating)

This alternative to making "popping boba" doesn't require any specialized ingredients, but is more like a lab experiment, not a snack. The fact that oil and water do not mix is a key to forming the gelatin into spheres.

You will need:
- 5 packets unflavored gelatin (or 2 tablespoons agar powder)
- distilled water
- vinegar
- food coloring
- very cold vegetable oil (put in freezer until cloudy but not frozen)
- several empty squeeze bottles (or basters or syringes)
- optional: strainer for rinsing off spheres
- optional: purple cabbage water for coloring-changing spheres

How to prepare:
1) Pour vegetable oil into a tall container and put into the freezer until it turns cloudy but is still liquid, not solid.

What to do:
1) Dissolve 5 packets gelatin powder (or 2 T. agar powder) into 1 cup (250 ml) hot water. Add 2 teaspoons (10 ml) vinegar. Heat and stir till dissolved.
FOR COLORING CHANGING SPHERES, do not add the vinegar and use 1 cup of purple cabbage water instead of plain water.
2) Pour into a few smaller containers and add drops of food coloring to each, so you have several colored gelatin mixes to work with. Allow to cool a bit, so that it is not dangerously hot. Do not allow to cool completely.
3) Pour each colored mix into an empty squeeze bottle (or whatever you are using).
4) Slowly squeeze out small amount of the mixture into the cold oil. The drops should form a marble-sized blob that will become round as it sinks in the cold oil. Allow these spheres to cool at the bottom of the oil for about 30 seconds, then pull them out. Rinse the spheres with water.
5) Try squeezing two colors out at the same time for multi-colored spheres.
6) The sphere can be allowed to dry on a plate overnight, and can then be rehydrated with water the next day.
7) For COLOR CHANGING spheres, drop them into vinegar or baking soda water to see them turn pink or blue/green.

ACTIVITY IDEA 6I COOKING ACTIVITY: TRY MAKING AN ASPIC

You might want to search the Internet for a few aspic recipes to find a blend of meat/egg/vegetables that sounds particularly appealing to you.

You will need:
- agar powder
- boiling water or broth
- your choice of meat, vegetable, egg, etc.
- a pan or jello mold

What to do:
1) Check the directions on your package of agar. Mine said to use 1 teaspoon (5 ml) agar powder for one cup (250 ml) of liquid.
2) Mix the agar and water/broth. Bring to a boil for one minute, then turn off heat and let it cool for a few minutes.
3) Pour warm agar into your dish and then add your layers of meat/egg/vegetables.
4) The aspic will harden even without chilling, but chilling can speed up the process. Once gelled, the agar will stay firm even without refrigeration (as we read in the chapter). However, I advise keeping it chilled if it needs to be stored overnight or longer.

ACTIVITY IDEA 6J   COOKING ACTIVITY: HOW FATS AFFECT A CAKE RECIPE

This activity explores the role of fats in baked goods such as cakes and cookies. You will be making small cakes using various types of fats. You will keep all other variables in the recipe the same, so the only difference between the cakes will be type of fat used. You can use the recipe given here, or you can use any recipe that suits you (gluten-free, sugar-free, egg-free, etc.) as long as you keep everything the same in each cake except for the type of fat. If using your own recipe, you will probably want to cut it in half (or even in thirds, if possible) so that you won't end up producing more cake than you can possibly eat. Usually, the limiting factor in reducing a recipe is the number of eggs, since it is hard to use half an egg. Therefore, if your recipe calls for three eggs, it will be easy to make a one-egg recipe and divide the measurements of the other ingredients by three. A two-egg recipe should be divided in half.

OPTIONAL: You might also want to make a cake that has no fat, and see what happens.

What to tell the students:

We've read about a number of fats in this book. Some are solid at room temperature, such as butter and shortening, and others, such as olive oil or vegetable oils, are liquids at room temperature. Fats are what give baked goods their enjoyable texture. They contribute to the stability of dough or batter so it can hold the air bubbles produced by the leavening agent. They also help to hold moisture, which makes the baked good more tender.

Professional bakers have a special term for how a baked good feels in your mouth as you chew. They called it **"mouthfeel."** *Pie crust gives a soft and flaky mouthfeel. Cookies have a tender and crumbly mouthfeel. Cake gives you a soft, fluffy mouthfeel. The type of fat used in a recipe contributes substantially to the mouthfeel of the final product. In this activity you will experiment with different types of fat to see how they affect the mouthfeel of a cake.*

You will need:
- a cake recipe that suits you, if you can't use the one listed below
- flour (can substitute gluten-free if you need to)
- a sweetener (if using a sugar substitute, figure out how much of your sweetener equals 1/3 cup (67 g) sugar)
- baking powder
- baking soda
- eggs (if you can't use eggs, consider a plant-based egg substitute)
- milk (can use dairy-free milk such as almond or soy)
- vegetable oil (any mild-tasting oil)
- butter
- shortening
- optional: another type of fat (or you can use an extra egg for your fat)
- small baking dishes, or cupcake molds and/or papers

Recipe (for a yield of small cakes):

1/4 cup (50 g) sugar   1 large egg
1 teaspoon (5 ml) vanilla   3/4 cup (96 g) flour
3/4 teaspoon (4 g) baking powder   1/4 cup (60 ml) milk
one of these fats:
    1/4 cup (58 g) butter
    1/4 cup (58 g) shortening
    1/4 cup (60 ml) vegetable oil (such as corn oil, safflower oil, sunflower oil, or mild-tasting olive oil)
    1 extra egg (or other type of fat if you want to try something like coconut oil)

What to do:
1) Set oven at 350 degrees F (175 degrees C).
2) Put the sugar and fat into a bowl and use an electric mixer, or hand whisk, to "cream" the fat and the sugar. Beat until fluffy, but do no over-beat. Note that the liquid oil won't "cream" in the same way, so don't over beat it.
3) Add the egg (or eggs if using an extra egg as your fat) and the vanilla and beat briefly.
4) Put the flour into a separate bowl and add the baking powder. Mix well. If you a have a sifter, sift the dry ingredients into the wet ingredients bowl. If not using a sifter, just make sure the baking powder is thoroughly mixed before adding. Blend the dry and wet ingredients well.
5) Add the milk and blend until thoroughly mixed.
6) Pour the batter into the dish (or cupcake molds) and make sure you mark them so you can keep track of which is which.
7) Bake until the top of each cake springs back when touched. The number of minutes will depend on the size of your cakes.
8) Sample each cake and compare the textures. Which type of fat made the texture you like best? Least?

ACTIVITY IDEA 6K   CARD GAME: "Thickening Memory Match"

You will need:
- one copy of this page, one copy of the next page, printed onto card stock if possible

How to prepare:
1) Copy these pattern pages. Cut the cards apart.
2) Let the players see all the cards first. Let them identify which cards go together and **spend a few minutes reviewing what happens in each thickening process**. (Sets are: gelatin, starch, whipped cream, butter, egg whites, egg yolks, and agar.
3) Shuffle cards very thoroughly and place them all face down, making a grid.

How to play:
1) Use standard "memory match" rules. Each player gets a chance to turn over three cards. If they match, they keep that set of three. If not, they return them face down. Play until all cards are gone. Player with most matches wins.

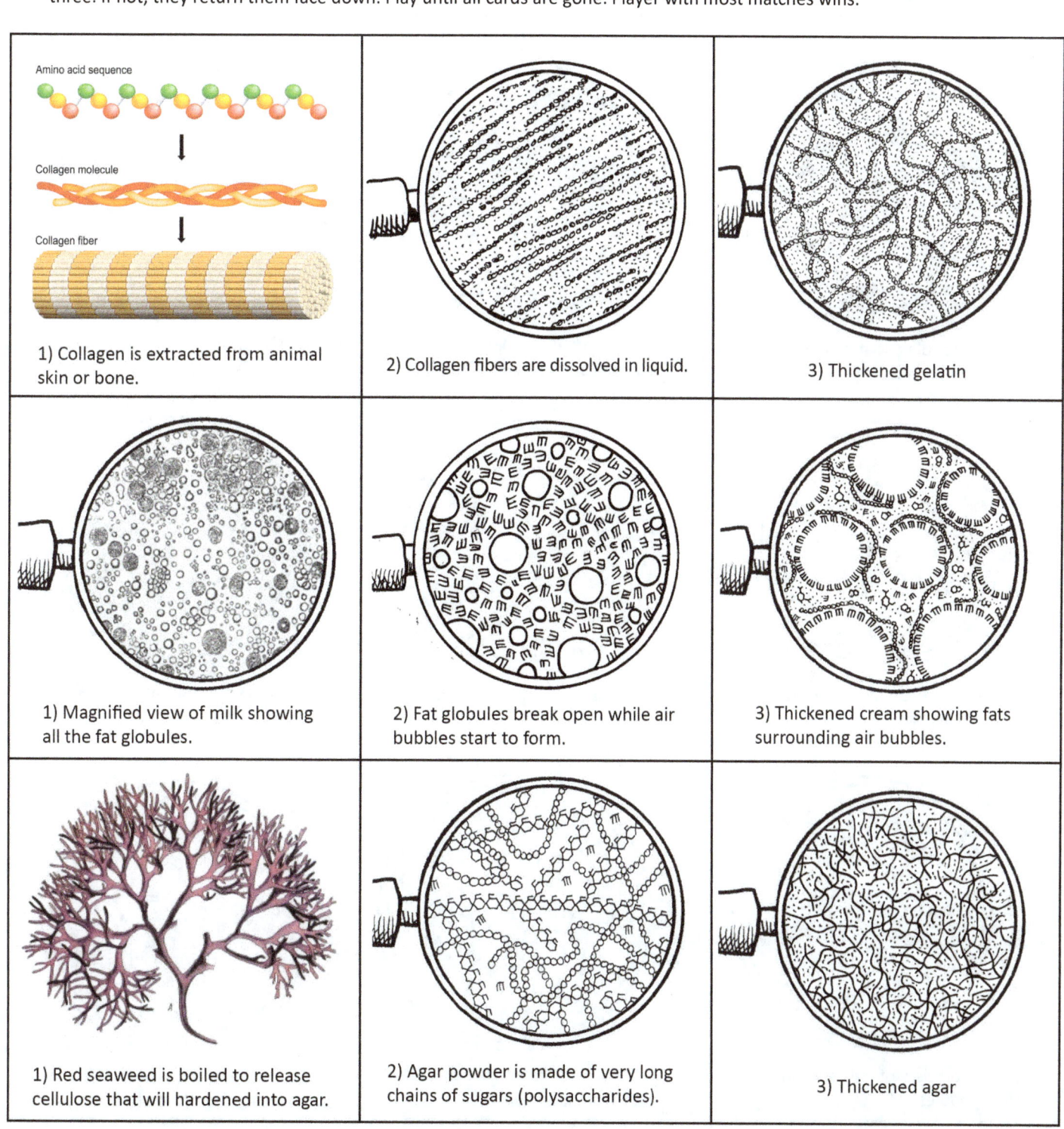

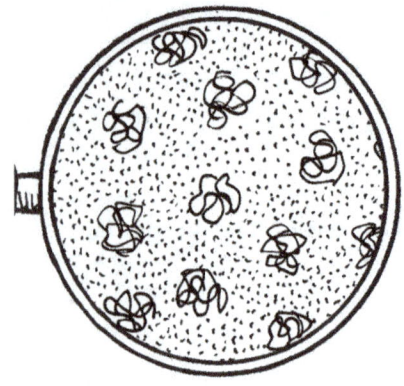

1) Egg white before whipping, with clumps of proteins.

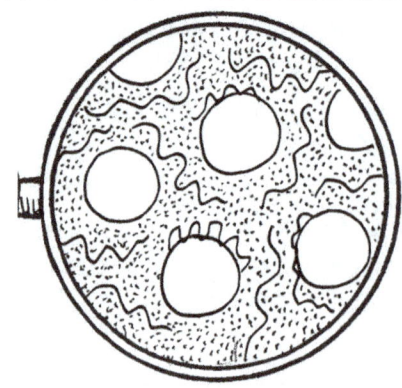

2) Egg white during whipping. Proteins unfold and air bubbles appear.

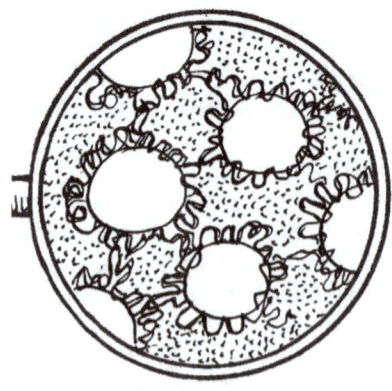

3) Proteins stick to the edges of the air bubbles and stabilize them.

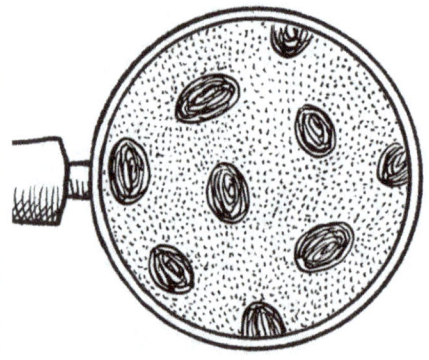

1) Starch granules floating in water. The granules contain amylose chains.

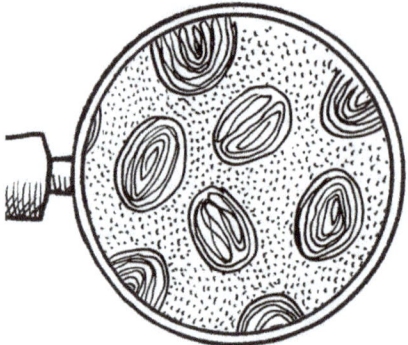

2) Starch granules swell up with water as the water becomes very warm.

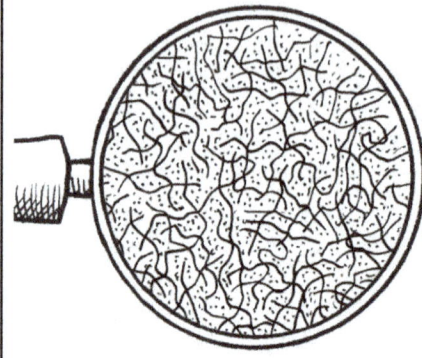

3) Starch granules explode and release amylose. Water can't flow very well.

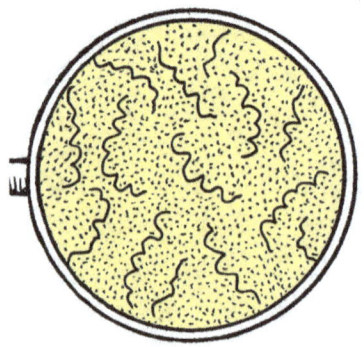

1) Heat causes the clumped proteins to unfold and become "denatured." Water flows freely around them.

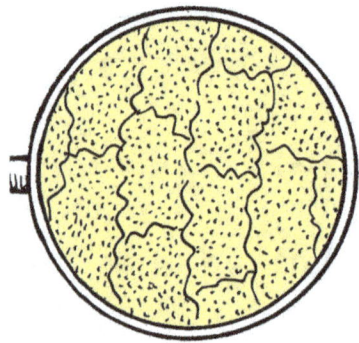

2) The proteins begin to coagulate and form new structures. Water is trapped in pockets and cannot flow.

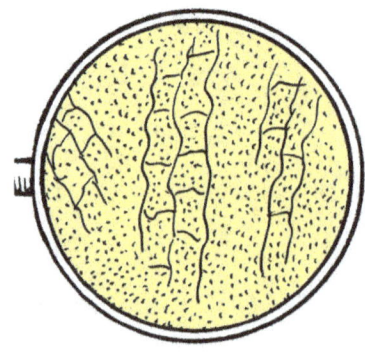

3) This is what happens when egg proteins are over-cooked. The structures shrink and water is released.

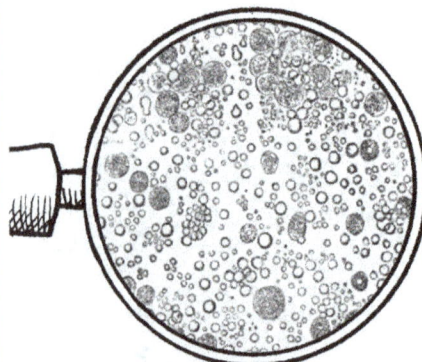

1) Magnified view of milk showing all the fat globules.

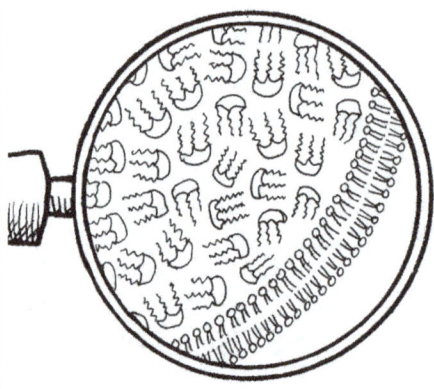

2) Triglyceride fat molecules must be released from globules by "churning."

3) Triglycerides are completely free, and they stick together to make butter.

ACTIVITY IDEA 6L   MOLECULE MAT for chapter 6

You will need:
- a copy of the Molecule Mat chapter 6 pattern page
- toothpicks
- the materials for the atoms

What to do:
1) Put your chosen materials inside the boxes on the left side of the page (or in small dishes if they won't fit inside the boxes). Toothpicks can be set in a dish, or simply in a pile, within the students' reach.
2) Let the students work on their own as much as possible.
3) The molecules may or may not all fit onto the page. Xylitol can be done separately, and the materials from the other molecules can be recycled.
4) Note that for xylitol, the students will have to "read" the diagram and figure out where the invisible C's and H's are.

ACTIVITY IDEA 6M   TASTE TESTING SUGAR SUBSTITUTES

NOTE: All artificial sweeteners are considered "safe" unless you are allergic to them (or have a very rare and unusual non-allergic sensitivity). While I would not recommend consuming artificial sweeteners every day for a very long time, tasting a sample in science class is perfectly harmless (barring any allergies). Some students could be overly concerned about tasting something "artificial" and might need reassurance that doing this scientific taste test will not cause them any harm. We need to remember that even natural sugar can be very harmful to our health, and is definitely a contributing factor to many chronic illnesses.

You will need:
- a variety of sugar substitutes (can also include one or more natural sweeteners such as monk fruit, stevia, or xylitol)
- sugar
- paper cups
- some packets of flavoring that contain only color and flavor (ex: Kool-Aid®) If you want to avoid artificial colors, you can use any type of flavoring that does not contain any sweetener
- marker or pen (if you want to use one cup per sample)
- NOTE: If you want to use a minimum number of cups, you can make pitchers of the drinks and provide each student with one cup that you will rinse and refill for each type of drink.

What to do:
1) Each student will need a set of small cups. (In my class I used five cups per student, one for sugar and four for the substitutes, but you can do more.) Or, if you want to avoid using a lot of cups, you can give one cup to each student and have everyone taste the same sample at the same time, emptying the cup (maybe even rinsing?) the cups between samples.
2) If using a cup per sample, label each cup with the name of one type of sweetener.
3) Mix pictures of "juice" solutions, with the same amount of flavoring but a different sweetener in each.
5) Pour the juice solutions into the labeled cups.
(Optional: Provide one cup with just the flavor and no sweetener at all.)
6) Have the students try each cup, sipping slowly and tasting carefully. Ask follow-up questions. Which drink was their least favorite? Did any of the samples produce an "after taste"? Did any of the samples taste bitter? Which sample came closest to giving the same effect as sugar? If you could no longer eat sugar, which substitute would you choose?

Splenda® = sucralose   Nutrasweet® = aspartame
Equal® = aspartame   Sweet N Low® = saccharin
Sugar Twin® = saccharin

ACTIVITY IDEA 6N   OPTIONAL: REVIEW/TEST

You can use the following pages (after the molecule mat) as a final exam, or just as a final review activity. Or skip it and just end with the dessert activities. If you decide to use it as an exam, and you want to have the students study ahead of time, tell them the best way to prepare is to read over the review questions at the end of each chapter, and to make sure they know the definitions of the words printed in ***bold italic.*** The questions are very basic and are mostly about the types of molecules.

Consider giving partial credit for answers that are spelled incorrectly. If you make each question worth two points, then it is very easy to give partial credit and calculate a final score based on 100 percent.

-----Caramelization molecules-----

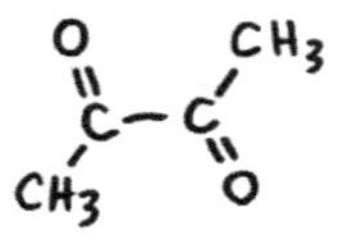

diacetyl

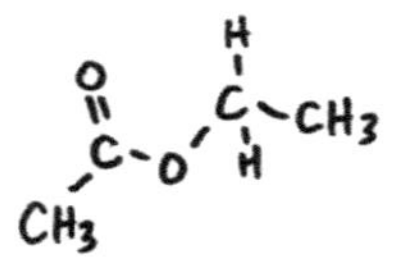

ethyl acetate

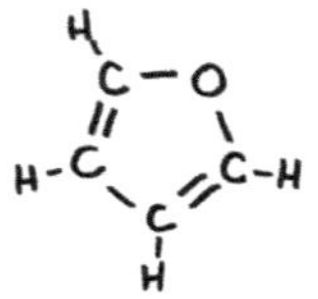

furan

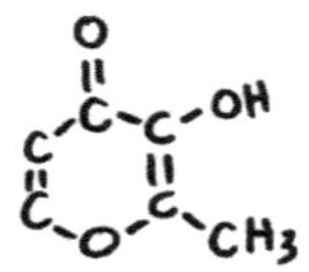

maltol

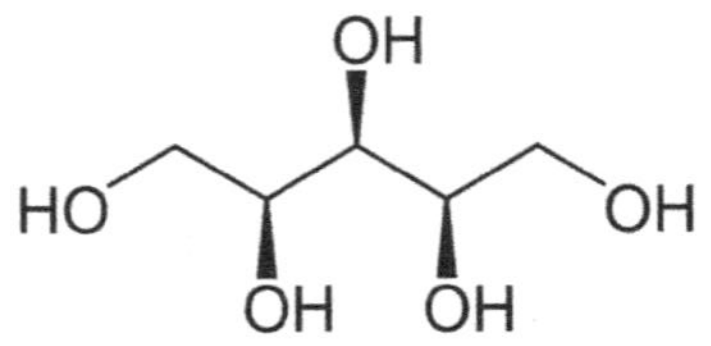

xylitol
(There are 5 carbons, and 7 hydrogens not shown. Remember, carbon always makes 4 bonds!)

H - hydrogen
O - oxygen
C - carbon

MOLECULE MAT Chapter 6

DISSECT YOUR DINNER   Final review/quiz   Name _____

1) What do you call the electrical attraction between the positive and negative sides of water molecules? _____ bonding

2) In salt water, which is the "solute," salt or water? _____

3) What do you call the things that enzymes bring together or tear apart? _____

4) Molecules that end in "-ose" are probably: a) fats   b) proteins   c) sugars   d) polar molecules   e) enzymes

5) What molecule is often used to "patch" the broken ends of molecules that have been cut apart by an enzyme? _____

6) Which one of these is NOT a factor when trying to put a lot of carbon dioxide into water (to carbonate it)?
a) cold temperature   b) high pressure   c) amount of sugar   d) large surface area

7) Which one of these would you be LEAST likely to find in a can containing a sweetened fizzy drink?
a) potassium benzoate   b) sodium acetate   c) phosphoric acid   d) caffeine

8) Where would you find the bacteria called *Lactobacillus*? _____

9) What number on the pH scale is neutral? _____

10) When you mix an acid and a base, you get water and a _____.   a) salt   b) sugar   c) crystal   d) liquid

11) The process of heating milk to kill bacteria is called _____.

12) The process of pushing milk through a sieve to make all the fat globules the same size is called _____.

13) A triglyceride's "tails" are made of _____ _____.

14) Proteins are made of individual units called _____ _____.

15) The primary protein in milk is called _____.

16) When a water molecule breaks apart, the parts are called: _____ ion, and _____ ion.
                                                                (OH-)   (H+)

17) Butter is yellow because it contains this molecule: _____

18) Bacteria produce lactic acid during the process of _____.

19) When an acid is put into milk, it will cause it to _____.

20) Yeast makes bread rise by producing bubbles of _____ _____.

21) When you make cheese, the watery stuff left over is called _____.

22) Amylose and amylopectin are types of _____.

23) When broken pieces of protein and sugar molecules join together (due to heat) this is called:
a) caramelization   b) gelatinization   c) Maillard reaction   d) curdling   e) denaturing

24) Where do pyruvate molecules come from?   a) glycine   b) triglycerides   c) glucose   d) acids

25) If your quick bread recipe calls for baking soda, what will be needed to make it work? (to make gas bubbles)
a) an acid   b) a base   c) water   d) milk   e) carbon dioxide

26) A substance that can hold on to both polar and non-polar molecules is called an _____.

27) Name the molecule in leaves that makes them green: _____

28) Cell membranes (in either plant or animal cells) are made of 2-legged molecules called:
a) diglycerides   b) phospholipids   c) carotenes   d) thylakoids   e) chlorophyll

29) What type of salad green has the most vitamin K, C, B2, B9, magnesium, calcium, and phosphorus? _____

30) Which plant pigment is responsible for red and blue colors?
a) anthocyanin   b) beta-carotene   c) chlorophyll   d) xanthophyll

Match the molecules with the names.

31) beta-carotene _____   32) chlorophyll _____   33) triglyceride _____   34) amino acid _____   35) glucose _____

A   B   C   D   E

36) Polyphenol chemicals are produced by plants as:
a) sources of nutrition   b) scents to attract birds   c) molecules that collect light   d) pesticides

37) Which vitamin cured scurvy?   a) vitamin A   b) one of the B vitamins   c) vitamin C   d) vitamin D

38) Which vitamin is associated with "pernicious anemia," has a cobalt atom in its chemical structure, makes red crystals when it is purified, and is found in meat, diary products, and shellfish (crabs, lobsters)?   a) B6   b) B12   c) C   d) D

39) Which of these vitamins is water-soluble (not fat-soluble)?   a) A   b) B1   c) K   d) D

40) You can't digest cellul*ase*. What living thing CAN digest it? _____

41) Which of these molecules is considered to be toxic?   a) chlorophyll   b) raffinose   c) capsaicin   d) solanine

42) Phytohemagglutinin is a toxin found in: a) beans   b) potatoes   c) red meat   d) starch

43) What is starch made of? strings of _____ molecules

44) Where do you find myofibrils, actin and myosin? _____

45) What is it called when heat "ruins" the shape of a protein and causes it to unfold? d_____

46) Which is more likely to be solid at room temperature?   a) saturated fats   b) unsaturated fats

47) "Trans" is Latin for "across." In trans fats, what is located across (on opposite sides of) the molecule?
a) carbon atoms   b) hydrogen atoms   c) oxygen atoms   d) double bonds

48) Which of these is NOT used to thicken desserts?
a) agar   b) pectin   c) collagen   d) carrageenan   e) eggs   f) starch   g) gluten

49) Sucrose can be separated into glucose and fructose by two of these. Which one is not used?   a) heat   b) cold   c) enzymes

50) Which of these was not discovered by accident?   a) aspartame   b) stevia   c) saccharin   d) sucralose

ANSWER KEY TO final review

1) hydrogen bonding
2) salt
3) substrates
4) c) sugars
5) water
6) c) amount of sugar
7) b) sodium acetate
8) milk
9) 7
10) a) salt
11) pasteurization
12) homogenization
13) fatty acids
14) amino acids
15) casein
16) hydroxide ion, hydrogen ion
17) beta-carotene
18) fermentation
19) curdle
20) carbon dioxide
21) whey
22) starch
23) c) Maillard reaction
24) c) glucose
25) a) an acid

26) emulsifier
27) chlorophyll
28) b) phospholipids
29) spinach
30) a) anthocyanin
31) D
32) E
33) A
34) C
35) B
36) d) pesticides
37) c) vitamin C
38) b) B12
39) b) B1
40) bacteria
41) d) solanine
42) a) beans
43) glucose
44) muscles
45) denaturing
46) a) saturated fats
47) b) hydrogen atoms
48) g) gluten
49) b) cold
50) b) stevia

ACTIVITY IDEA 6O   A REVIEW QUIZ GAME ("Jeopardy" style)

You can modify this game for individual use, but these instructions will be for use in a group.

You will need: copies of the pattern pages, scissors, tape, pencils and paper for keeping score, possibly a clock or timer to keep track of 15-second turns

1) Make one copy of each title page (AMINO ACIDS/ PROTEINS/ FATS, etc.). Make 6 copies of the number page (100, 200, etc.)
TIP: To make your game board look nice, consider printing on colored paper.

2) Cut apart all the titles and numbers.

3) A wide variety of topics are given. Choose 6 of the topics for your first game. There are enough topics to play 3 games and still have some left over. (For a shorter game, choose less than 6 topics.) You can use them in any order, and you don't have to use all of them.

4) Tape the titles you have chosen to a blank wall (or place on a table if a wall is not possible). Under each topic, tape (lightly) the numbers 100 to 500.

5) Divide your players into two or three groups. In my experience, it works much better to simply have the teams take turns at answering questions, rather than making it a contest of who is fastest to respond (no "buzzers").

6) Have the players on each team take turns answering. However, I recommend allowing the person who is answering to get advice from their team members. This alleviates a lot of nervousness, and prevents a player from embarrassment in front of their peers if they don't have a guess. The spokesperson is the one who will choose which topic and which number. Follow Jeopardy format, for example, "Proteins for 300."

7) Pull down the number that was just called and remove it from the board. Then find the corresponding question in the list of questions. Read the question out loud and allow a short time for answering. I recommend limiting the time to about 15 seconds. It would help if you have someone who can watch the clock and ring a bell or something when 15 seconds has elapsed. If that team can't come up with an answer in 15 seconds, the question will go over to the other team. If they miss, it is up to you whether to allow the first team to have another go at it.

8) Keep score as points are earned. Add up the totals after all the questions have been used.

9) Play again, if you can, with different topics.

ACTIVITY IDEA 6P   A GROCERY STORE SCAVENGER HUNT

You will need: copies of the scavenger hunt pages (following the quiz game pattern pages), pencils, possibly clipboards

Allow at least an hour for this scavenger hunt. Even an hour probably won't let everyone find everything unless your grocery store is small. If you are limited on time, you can also cross out one or more categories. You might want to spend a few minutes reading it over before starting the hunt, making sure the students understand what they are looking for.

QUESTIONS FOR THE QUIZ GAME

AMINO ACIDS
100: This is the smallest amino acid. (glycine)
200: What type of atom is at the center of an amino acid? (carbon)
    (HINT: It is sometimes called the alpha ___.)
300: An amino acid has a carboxyl group (COOH) on one side, and what group on the other? (amine)
400: Cysteine is the amino acid that is famous for forming cross bridges. (We saw these bridges between gliadin and glutenin when they formed gluten.) The key to cysteine's ability to form these bridges is what element? Hint: This element is also what makes many things stinky. (sulfur)
500: Name another amino acid besides glycine and cysteine. (Mentioned in this book: alanine, lysine, threonine, glutamine (also known as glutamate or glutamic acid), proline, methionine.)

PROTEINS
100: Proteins are made of individual units called ___. (amino acids)
200: Name the major protein found in milk. (casein)
300: Where do you find a peptide bond? (between amino acids)
400: What element is found in proteins but not in sugars or fats? (nitrogen, N)
500: When proteins unfold and lose their shape as they are heated we say they become ___. (denatured)

FATS
100: Saturated fats have their maximum number of what? (hydrogen atoms)
200: What do you call a glycerin molecule with three fatty acids hanging from it? (triglyceride)   300: What type of fatty acids have hydrogen atoms missing at the third carbon from the end of the chain? (omega-3's)
400: Which is more likely to be solid at room temperature-- saturated or unsaturated fats? (saturated)
500: This molecule is found in membranes and has two fatty acid tails (phospholipid)

CARBOHYDRATES
100: Carbohydrates are primarily made of what two elements? (carbon and hydrogen)
200: A long string of glucose molecules that our bodies cannot digest: ___. (cellulose)
300: Potatoes and sweet potatoes are classified as this plant part. (tubers)
400: What organelle inside plant cells stores starch molecules? (amyloplasts)
500: Where do plants get the carbon atoms that end up in the starch they make? (carbon dioxide in air)

SUGARS
100: Name a monosaccharide. (glucose, fructose or galactose)
200: A long string of glucose molecules that our bodies can easily digest. (starch)
300: Name a disaccharide. (sucrose, lactose)
400: How many carbon atoms are in simple sugars like glucose and fructose? (6)
500: What is the correct name for the process of splitting a glucose molecule in half? (glycolysis)

BEVERAGES
1- Which temperature, hot, warm, or cold, is best for forcing carbon dioxide to dissolve into water? (cold)
2- Is milk a solution, colloid, or mixture? (colloid)
3- Is cola a solution, colloid, or mixture? (solution)
3- Name a preservative you might find in a can of juice or soda? (possible answers: sodium benzoate, potassium benzoate, vitamin C, sugar)
5- What is the pH of pure water? (7)

CHEMISTRY
100: Name the two parts of a hydrogen atom. (proton and electron)
200: A molecule with a positive side and a negative side is said to be ___. (polar)
300: The attraction between water molecules is called: ___. (hydrogen bonding)
400: When a water molecule loses one of its hydrogens, the result is this ion. (hydroxide)
500: Name the solvent in salt water. (water) (salt is the solute)

ENZYMES
100: What enzyme can break apart sucrose molecules? (sucrase)
200: The molecules that enzymes either join together or break apart are called: ___. (substrates) (This was voted the most forgettable word in chapter 1.)
300: Sometimes a tiny molecule is needed for an enzyme to be the right shape. What are these tiny helper molecules called? (coenzymes)
400: Your body doesn't make the enzyme needed to break apart cellulose. What type of organism does make this enzyme? (bacteria)
500: What are enzymes themselves made of? (mostly protein)

MILK
100: What kind of sugar is found in milk? (lactose)
200: What do you call the process where all the bacteria in milk are killed? (pasteurization)
300: What do you call the process where all the fat globules in milk go through a strainer to make them the same size? (homogenization)
400: What is the primary protein found in milk? (casein)
500: Negatively charged sugar molecules on the edges of protein micelles in milk provide repelling forces that keep the protein micelles away from each other. To curdle milk, you must overcome these negative charges. Can you name one of the two ways to deal with these negative sugars? (Give them positive protons from acids, or cut them off using enzyme scissors.)

CHEESE
100: What is the correct name for the watery part of cottage cheese? (whey)
200: Which type of cheese is made by bacteria that produce a lot of carbon dioxide gas? (Swiss cheese)
300: Name a type of cheese (other than cottage cheese) that does NOT require microorganisms. (mozzarella, cream cheese, "American" cheese, Velveeta brand cheese)
400: This is the most popular cheese in the world. It originally came from a small town in England. (cheddar)
500: Casu Marzu cheese is made on the island of Sardinia. This cheese is famous for having living creatures crawling around in the cheese while you eat it. The tiny creatures can jump off the cheese so you have to hold your hand over the cheese as you eat. What are these living creatures? (fly maggots)

BREAD
100: Bread contains both protein and starch. What is starch made of? (long strings of glucose molecules)
200: Name one of the waste products that yeast cells make. (carbon dioxide, ethanol)
300: Part of what makes baking bread smell so good is the formation of strange molecules that are made of broken protein and sugar molecules that are randomly combined together. What are these types of molecules called? (Maillard)
400: What is the auto-immune disease where people have a serious reaction to gluten? (celiac)
500: Name one of the two proteins that make gluten. (gliadin or glutenin)

LEAVENING
100: When yeast makes dough rise, it is because of this gas made by the yeast (carbon dioxide)
200: To make a quick bread rise, you can use baking soda and a mild acid and they will react to form bubbles of___. (carbon dioxide)
300: Baking soda's chemical formula is $NaHCO_3$. What does the Na stand for? (sodium)
400: Which of these has a dry, powdered acid added to it? baking soda or baking powder? (powder)
500: Where was sodium bicarbonate, $NaHCO_3$ first discovered? a) in ocean water   b) in rocks   c) in plants   d) in dirt   e) in a lab   (in rocks)

VITAMINS
100: A deficiency of this vitamin causes scurvy. (C)
200: Riboflavin and niacin belong to this group of vitamins. (B)
300: This vitamin, also called cobalamin, contains the element cobalt and is found almost exclusively in foods that come from animals. (B12)
400: This vitamin is essential for the blood clotting process. (K)
500: This vitamin is produced in our skin cells when exposed to UV light. (D)

MICROBES
100: What is the collective term for all the microorganisms that live in your gut? (microbiome)
200: What type of microorganism can reproduce by budding? (yeast)
300: Limburger cheese is made with bacteria that can also be found on what body part? (feet)
400: What microorganism was genetically engineered to produce the leghemoglobin molecules used in plant-based meats? (yeast cells)
500: The process where bacteria turn pyruvate molecules into lactic acid is called: ____. (fermentation)

PHYTONUTRIENTS
100: Vitamin A is made from which phytonutrient? (beta-carotene)
200: Which phytochemical is the primary participant in photosynthesis? (chlorophyll)
300: Anthocyanin can turn red or blue depending on what? (pH)
400: What polyphenol chemical is found in great abundance in tea leaves and gives tea its tangy flavor? (tannins)
500: Resveratrol is a phytonutrient found in dark red or purple fruits. Because it helps to protect your body against dangerous oxygen atoms, it is classified as an ___. (antioxidant)

BUTTER
100: Butter is yellow because it contains this molecule. (beta-carotene)
200: When you churn butter, there is a watery substance leftover. What is it called? (buttermilk)
300: The individual fat molecules in butter are called: ____. (triglycerides)
400: What does churning do to the fat globules in milk? (breaks them open releasing triglycerides)
500: Name an animal that produces milk that will make white butter. (sheep, goat) (The butter is white because the animal's cells chop beta-carotenes in half and store them as retinol, which doesn't reflect as much yellow light.

PLANTS
100: Which one of these is a true vegetable? tomato, squash, corn, peas, beans, lettuce, cucumber
   (Lettuce, because anything with a seed is technically a fruit.)
200: Name two anatomical parts of a seed.
   (plumules, cotyledon, seed leaves, seed coat, funiculus, embryo, endosperm)
300: Bean pods are technically what part of the plant? (ovaries)
400: What plant is the basis for plant-based meat (for both protein and leghemoglobin)? (soybean)
500: Pepper plants produce a chemical that stimulates the hot sensors on your tongue.
   What is this chemical called? (capsaicin)

THICKENING
100: What type of molecule collects around the air bubbles in meringue? starch, protein, or fat? (protein)
200: Unlike gelatin, agar is stable at room temperature. What is agar made from? (seaweed)
300: To make gelatin, you extract this from bones or skin. (collagen)
400: What percentage of fat do you need in cream in order for it to thicken when whipped?
   3%   12%   22%   35%   100%   (35%)
500: What do you call a savory, not sweet, gelatin that contains meat and/or vegetables? (aspic)

DESSERTS
100: When sucrose is heated, the heat separates sucrose into ____ and ____. (glucose and fructose)
200: Theobromine, the molecule that gives chocolate its distinct taste is very similar to what molecule that we find in coffee and tea? (caffeine)
300: Cacao is a word from what ancient culture? (Mayans)
400: Why is cornstarch or tricalcium phosphate added to powdered sugar?
   (to absorb moisture and keep the sugar powdery and fluffy)
500: This is a flexible blend of gelatin and sugar that can be used to sculpt edible shapes to decorate cakes and cookies. (fondant)

TOXINS
100: Solanine, a toxin in potatoes, is easy to spot because it turns them ____. (green)
200: Theobromine, the chemical that makes chocolate taste like chocolate, is manufactured by the cacao plants to do what? (kill bugs, an insecticide)
300: Raffinose, the chemical in beans that our bodies can't digest, is made of three ____. (sugars)
400: The toxic chemical in beans, phytohemagglutinin, can be neutralized by doing what? (cooking)
500: What family of toxic plants includes potatoes, tomatoes, eggplants, and peppers? (nightshades)

AMINO ACIDS

PROTEINS

FATS

SUGARS

CARBOHYDRATES

BEVERAGES

CHEMISTRY

ENZYMES

MILK

CHEESE

BREAD

LEAVENING

VITAMINS

MICROBES

PHYTONUTRIENTS

BUTTER

PLANTS

THICKENING

DESSERTS

TOXINS

100

200

300

400

500

Grocery Store Scavenger Hunt

Remember to list the % Daily Value per serving (or the number of grams, or mg) per serving, so that you can compare your results to the results of the other participants.

PRODUCE:

List all the purple fruits or vegetables can you find. (Purple, not blue or red!) _____
_____

List any yellow fruits or vegetables you find: _____
_____

List as many types of salad greens as you can find: (Must be specific-- "spring mix" does not count.)
_____
_____
_____

List fruits grown outside of your country_____
_____

Can you find something locally grown? _____

FROZEN FOODS:

"Baby" vegetables: _____

Ice cream that has more than one flavor in the carton (ex: both chocolate and vanilla): _____
_____

Company that makes a pizza that does NOT have tomato sauce:_____

Types of fruit pies you can find: _____

MEAT:

Preserved meat with the least sodium % per serving: _____

Preserved meat with the highest sodium % per serving: _____

A preserved meat that has no preservatives (no nitrates or nitrites): _____

Heaviest package poultry (bird) meat: _____

PLANT-BASED MEAT:

How many companies are making plant-based meats? _____

BAKING

List types of liquid cooking oils you can find: _____
_____
_____

List solid fats designed for cooking: _____
_____

Spices beginning with the letter C: _____
_____

List all the low-calorie sugar substitutes you can find: (chemical name, not brand name, so read labels):
_____
_____

List gluten-free flours you can find (no wheat): _____
_____

Types of gluten-free pasta: _____
_____

CANNED GOODS

Soup with no sodium? _____

Soup with highest about of sodium: _____

Types of beans (just plain beans, not baked beans or chili): _____
_____

Canned fruit with no added sugar (what type of fruit?): _____

CEREAL AND SNACKS:

Boxed breakfast cereal with least amount of sugar per serving: _____

Boxed breakfast cereal with highest amount of sugar per serving: _____

Breakfast cereal with highest amount of iron per serving: _____

"Chip" with the least sodium per serving: _____

Pretzel product with highest amount of sodium per serving: _____

How many types of peanut butter substitutes can you can find? _____
_____

A brand of peanut butter that has no salt, sugar or added fats: _____

Cookie with highest number of calories per serving: _____

DAIRY:

Dairy product with the highest number of grams of protein per serving: _____

Dairy product with highest amount of fat per serving? _____

How many alternative "milks" can you find? _____
_____

Chocolate milk with highest amount of sugar per serving: _____

Types of cheese with blue/green mold: _____

A brand of cheese made in (and imported from) France: _____

CONDIMENTS

Salad dressing with least amount of sugar per serving: _____

Pickles with the least amount of sugar: _____

Ketchup with the highest amount of sugar: _____

BEVERAGES

A sports drink with no food coloring: _____

Carbonated drink (soda pop) with highest amount of sugar per serving _____

ADDITIVES

(Just put a check mark if you find it. You'll have to read labels to find them.)

_____ Xanthum gum (used for thickening, found in many sauces, puddings, ice creams)

_____ Carrageenan (used for thickening, found in puddlings, ice cream, and frozen desserts)

_____ MSG (monosodium glutamate) often found in soups and in Asian food

_____ Sodium benzoate (often found in drinks and condiments)

_____ Calcium propionate (often found in bread products)

_____ BHT or BHA (found in many cereal, snack, and bread products)

_____ A colorful snack that has zero artificial colors or flavors

OPTIONAL NUTRITION WORKSHEETS

If you are working with older students (especially those looking for high school credit) you might want to consider using these "scavenger hunts" to provide additional credit hours.

You can use one or two worksheets per chapter as you go along, or you can use them at the end, as additional homework, if you want to stretch the unit to fill a semester.

For a class, another option would be to bring in a selection of food containers to your classroom and do these worksheets in class.

If you are working with a class and need to fill additional class time, you might want to consider having some special guests come and speak to your group, such as a professional nutritionist (in private practice, or from a local school or hospital), a specialty cook, an organic gardener, a food chemist, or even someone who likes to cook food that is very different from what the students are used to (some type of ethnic food).

You might also consider assigning the students to research various diets (vegetarian, vegan, paleo, keto, etc.)

Research Challenge In Your Kitchen: Finding Sodium

This can be done at home, in your kitchen or pantry, but you can also do it during a trip to a grocery store (food market).

Salt (sodium chloride) is added to most food products, but nutrition labels only list "sodium" not "sodium chloride." If the food contains MSG (monosodium glutamate) this would be another source of sodium.

In the USA, the amount of each nutrient is listed as "% Daily Value." This means that if you add up numbers for all the food you eat in a day, the total should not be over 100%. Sometimes food companies also add the exact measurement, such as grams or milligrams. The recommended number of grams or milligrams can vary widely between foods, so the % Daily Value helps to keep the comparisons simple. If your country does not list % Daily Value, just list the measurements you find on the package. (The recommend daily limit for sodium is usually somewhere around 2,300 mg, which is about a teaspoon of salt.)

Compare three condiments that you have in your refrigerator. (ketchup, mustard, mayonnaise, pickle relish, hot sauce, etc.)

condiment type			
% Daily Sodium			

Compare three cereal/grain products you might eat for breakfast. (cereal, oatmeal, toast, granola, breakfast bar, etc.)

breakfast item			
% Daily Sodium			

Sometimes fruits and vegetables have labels and sometimes they don't. If you buy pre-bagged fruits and vegetables, try to find three bags with labels and compare them. If you grow your own or buy from a farm market and don't have any labels available, use a search engine on the Internet to find some. (Use key words such as "sodium content of apples.")

fruit or vegetable			
% Daily Sodium			

Find the sodium values for at least two preserved meats (bacon, sausage, pepperoni, ham, or any "lunch meat"). They can be fresh, frozen, or in a can. If you don't ever buy preserved meats, you can either use the Internet or do this research while at a grocery store.

type of preserved meat			
% Daily Sodium			

Salt is used in most cheese-making processes. Fresh cheese that are not "aged" tend to have less sodium. Try to find three different types of cheese and compare their sodium content. You can also use "processed" cheeses or cheese spreads.

type of cheese			
% Daily Sodium			

Find the sodium values for three beverages (something you drink).

beverage			
% Daily Sodium			

How much sodium is in your sweets? Find the sodium content for three things that you might eat for dessert or as a sweet snack.

dessert or sweet snack			
% Daily Sodium			

Which result was most surprising? _____ Which was least surprising? _____

Research Challenge In Your Kitchen: Finding Preservatives

This can be done at home, in your kitchen or pantry, but you can also do it during a trip to a grocery store (food market).

Preservatives are used not only to keep food from spoiling (due to bacteria or micro-fungi) but also to keep food looking, smelling, feeling, and tasting as fresh as possible. In an ideal world, we'd be able to have fresh food all the time without the need for any preservatives, but in the real world many people have limited access to fresh food, especially during winter and spring. Not all preservatives are manufactured; some are natural. Ascorbic acid (vitamin C), citric acid, tocopherols (vitamin E), salt, and sugar are examples of natural preservatives.

Preservative chemicals are always thoroughly tested before they are listed as "Generally Recognized As Safe" (GRAS). Generally safe does not, however, rule out the possibility that someone might have a bad reaction to it. Some people are either allergic or sensitive to certain preservatives and must avoid them. Even with GRAS chemicals, sometimes problems arise when they are exposed to high heat. Nitrates and nitrites, for example, can be modified by heat to form new chemicals that are not quite as safe. The risk is still small enough that most people go ahead and grill their meat instead of cooking it slowly at low temps.

Find 15 edible items that have one or more preservatives in them. Here is a list of common preservatives that you might find:

Ascorbic acid: (Vitamin C) Found in many beverages as well as fruit, or fruit-flavored, products.
BHA and BHT: (Butylated Hydroxyanisole and Butylated Hydroxytoluene) Like EDTA, these are primarily used to prevent food molecules from combining with oxygen (oxidation). BHA is a waxy substance and BHT is a powder. BHT is commonly found in many cereals, breads, and other boxed products. BHA has been under investigation for safety and might be harder to find.
Citric acid: Naturally occurring in citrus fruits. It is often added to beverages, jams, jellies, and fruit-flavored products.
Benzoic acid or Sodium benzoate: Used to preserve acidic foods such as pickles and fruits; also often found in carbonated beverages.
Calcium propionate: Slows down growth of many microbes. Found in cereals, breads, beverages, processed meats, cheese, and more.
EDTA: *(ethylene-diamine-tetra-acetic acid)* Found in canned goods, sauces, and sodas. It prevents the food molecules from combining with oxygen. Oxidation causes unpleasant tastes and can lead to foods going rancid.
Nitrates and nitrites: (Ex: sodium nitrate) Found most often in preserved meats such as ham, sausage, bacon, and lunch meats. Besides preventing spoilage, these chemicals also keep meats from losing their color.
Potassium sorbate: Discovered in 1850s from berries of mountain ash tree. Widely used in beverages, syrups, sauces and more.
Sulfites: (sulfur dioxide, sodium sulfite, sodium bisulfite, potassium bisulfite, sodium metabisulfite and potassium metabisulfite) The primary use is in dried fruits, where they prevent spoiling and browning. Up to 1% of the population is sensitive to sulfites.
Tocopherols: (Vitamin E) Very widely used, since it is natural. Found in cereals, breads, snacks, vegetable oils, and meats.

list your foods in this column	ascorbic acid (vit C)	BHA / BHT	Citric acid	Benzoic acid sod. benzoate	Calcium propionate	EDTA	nitrates or nitrites	Potassium sorbate	sulfites	tocopherols (vit E)
maple syrup				X				X		

Research Challenge: Serving Size, Protein, and Looking for Labels

This can be done at home, in your kitchen or pantry, but you can also do it during a trip to a grocery store (food market).

SERVING SIZE

One of the most important elements of a nutrition label is the serving size. All of the nutritional facts are based on the serving size. Serving sizes tend to be modest (not terribly large). For example, a typical serving size for nuts is 1/4 cup. For chips, it's between 10-20 chips, depending on size (less chips if they are large). For cookies, it will be either one or two cookies, depending on size. Most teenagers will eat at least two "servings" of a snack, if not more. Thus, if you want to calculate how many nutrients you are actually taking in, you must carefully look at serving sizes and compare them to the actual amounts you are eating. You might need to double (or triple?) the numbers.

Find two food products that have serving sizes that you think are too small. (You think most people eat more than this at one sitting.)
1) _____
2) _____

Find two products that have serving sizes that you think are about right.
1) _____
2) _____

Find one food product that has a serving size that you think is too large. (Most people would eat less than this at one sitting.)
1) _____

PROTEIN

First, how much protein do you need each day? A minimum of .8 grams per kilogram of body weight (.36 grams per pound)
Babies: 10 grams/day   School-age kids: 20-30 grams/day   Teenage boys: 50-55 grams/day   Teenage girls: 45 grams/day
Most people who live in developed countries eat more than they need.

Animal-based foods (meat, fish, eggs, dairy)
NOTE: If no label, consult Internet.

name of food:	grams protein/serving

Plant-based foods
(beans, nuts, fruits, vegetables, cereals, snacks)

name of food:	grams protein/serving

LOOKING FOR LABELS   (If you don't have one of these in your house, that's okay)

Do spices have nutrition labels? _____

Does salt have a nutrition label? _____ (the large size that you buy for baking, not a tiny salt shaker)

Does baking powder (or baking soda) have a nutrition label? _____

What about corn starch? _____   Vinegar? _____   Flour? _____   Bag of sugar? _____   Bouillon? _____

Coffee? (freeze-dried or ground coffee, not bottled drinks) _____   Tea bags? (real tea or herbal) _____

Bottled water? _____

Research Challenge In Your Kitchen: Fat and Cholesterol

This can be done at home, in your kitchen or pantry, but you can also do it during a trip to a grocery store (food market).

Fat content is a required listing on all foods. Fat is broken down into four categories: **saturated fats** (where all the carbon atoms are completely surrounded by hydrogen atoms), **monounsaturated fats** (where one hydrogen atom is missing, causing one kink in the chain), **polyunsaturated fats** (where many hydrogen atoms are missing, causing many kinks), and **trans fats** (where the hydrogen atoms attached to the carbon double bond C=C are on opposite sides, preventing a kink from forming). You will probably have trouble finding a product that contains trans fat because after studies showed the dangers of eating too much trans fat, many governments passed laws that forced food companies to find ways of eliminating trans fats from their products. Trans fats are created when unsaturated fats have hydrogen atoms added to them artificially by chemical processes. The reason that food companies were saturating the fats was to make them more "shelf stable." Products that contain unsaturated fats go stale or rancid more quickly.

Cholesterol, though it might be considered a type of fat, gets its own listing. Many people are concerned about cholesterol levels in their body and don't want to add a lot of extra cholesterol by eating foods that are high in this type of fat.

Compare three dairy products (milk, cheese, sour cream, ice cream, yogurt, etc.) You can use dairy substitues (soy, almond, etc.):

Dairy product:			
Total fat			
Saturated fat			
Trans fat			
Polyunsaturated fat			
Monounsaturated fat			

Choose three of the following: meat (fresh, canned or processed), egg, peanuts, cashews, tree nuts of any type:

High protein foods:			
Total fat			
Saturated fat			
Trans fat			
Polyunsaturated fat			
Monounsaturated fat			

Compare three snack foods:

Snack product:			
Total fat			
Saturated fat			
Trans fat			
Polyunsaturated fat			
Monounsaturated fat			

Compare three fruits or vegetables:

Fruit/veg. (one serving):			
Total fat			
Saturated fat			
Trans fat			
Polyunsaturated fat			
Monounsaturated fat			

Research Challenge In Your Kitchen: Finding Vitamins and Minerals

This can be done at home, in your kitchen or pantry, but you can also do it during a trip to a grocery store (food market).

Food companies (in the USA) are not required to give the consumer a complete list of vitamins and minerals. They are legally required to list iron, calcium, potassium and vitamin D. However, for some products, it is an advantage for the food company to furnish a more complete list. This is especially true for breakfast cereals. Perhaps studies show that people read labels more often at breakfast than at any other meal? (Do you stare at the box as you eat your cereal?)

Your checklist includes some of the B vitamins. In this group of vitamins, some of them are more commonly known by their names and some are more commonly listed by their number. So we have thiamine (B1), riboflavin (B2), niacin (B3), pantothenic acid (B5), pyridoxine (B6), and cobalamin (B12). For some reason, B6 and B12 are almost never called by their names, but thiamine, riboflavin, niacin and pantothenic acid are almost never called by their numbers. We've had to abbreviate the names here in this chart to make them all fit on one line.

Compare four cereals or other breakfast products (anything you can find that has a good list of vitamins).
Write the % Daily Value in each column. Sorry the spaces are so small!

breakfast items:	vit D	Ca	Fe	pot.	vit A	thia-mine	ribo-flavin	niacin	B5 (panto)	B6	folate	B12	Phos.	Mag.	Zinc

Compare three dairy products (or dairy substitutes). If the package does not give you all of these vitamins you can leave them blank.

dairy items:	vit D	Ca	Fe	pot.	vit A	thia-mine	ribo-flavin	niacin	B5 (panto)	B6	folate	B12	Phos.	Mag.	Zinc

Compare three snack products (any type, either natural or processed foods—whatever you snack on).

snack items:	vit D	Ca	Fe	pot.	vit A	thia-mine	ribo-flavin	niacin	B5 (panto)	B6	folate	B12	Phos.	Mag.	Zinc

For this chart, look around a bit and find four items that you think have an interesting list.

your items:	vit D	Ca	Fe	pot.	vit A	thia-mine	ribo-flavin	niacin	B5 (panto)	B6	folate	B12	Phos.	Mag.	Zinc

Research Challenge In Your Kitchen: Finding Sugar

This can be done at home, in your kitchen or pantry, but you can also do it during a trip to a grocery store (food market).

Some natural foods, such as fruits, have relatively high amounts of glucose, fructose, and other sugars (ending with "-ose.") These naturally occurring sugars are less of a concern in our diet than "added sugar." Natural sugars in plants are always found in close association with other molecules that decrease the negative effects of sugar. The problem comes when we remove the sugars from the plants, concentrate them, then add them to food, giving us a dose of pure sugar far higher than we would get from simply eating the plant. "Added sugars" (sucrose, glucose, fructose, dextrose, and others) are found not only in places where we expect them, such desserts and snacks, but also in places where we don't expect them, such as "healthy" frozen dinners, condiments, and soups. In this activity, we'll be look for both total amount of sugar (which includes both natural and added sugars) and just "added sugar." If you live in a part of the world where foods don't have both types of sugar labeled, just use the information you have.

In this activity, we'll be asking for both grams of sugar and % Daily Value. For sugar, especially, it is good to have an idea of how much a gram is and how many you need. For males, it is recommended to keep your "added sugars" under 36 grams. For females, it's less than 24 grams (due to smaller (on average) body size and muscle mass).

Compare three snack products that you often eat. (Candy, cookies, chips, granola bars, nuts, etc.)

Snack product			
Total sugars in grams			
Added sugars in grams			
Total sugars % Daily Value			

See if you can find this information for some fruits or vegetables. They can be fresh, frozen, or canned.

Fruit or vegetable				
Total sugars in grams				
Added sugars in grams				
Total sugars % Daily Value				

Find sugar information for milk (or a milk substitute such as soy or almond milk) and two other beverages (not water).

Dairy product	milk /milk substitute		
Total sugars in grams			
Added sugars in grams			
Total sugars % Daily Value			

How much sugar is in each of these?

	slice of cheese	slice of bread	an egg	can of soup
Total sugars in grams				
Added sugars in grams				
Total sugars % Daily Value				

Find the sugar information for various condiments (ketchup, mustard, mayonnaise, or salad dressings)

condiment or dressing				
Total sugars in grams				
Added sugars in grams				
Total sugars % Daily Value				

The following blank chart is useful if you want to have your students keep track of everything they eat for a day and compare what they eat to the recommended daily allowances.

Use one chart per day; so if you want them to log three days, give them three charts.

To make this task easier, the students can record percentages instead of actual measurements. They can then just add up the numbers in each column and record the total. It will be very easy to see how much above or below the recommended amount their intake was. If their body size is much less than average adult size, their totals should be less than 100.

Name _____   Date _____

Write in the name of each food you ate and the number of servings Use the food labels to fill in how much of each nutritional category that food provided. Don't forget that if you ate 2 servings of that food, your nutritional numbers will be doubled in each column! Add up the totals when you are finished. (Phytonutrients are the bright colors found in fruits and vegetables. For total phytos, add up the number of "Yes" answers.)

# of servings	NAME OF FOOD	Calories	Total fat	sat. fat	trans fat	Cholesterol	Sodium	Total carb	sugar	fiber	Protein	Vit. A	Vit. C	Calcium	Iron	Thiamin	Riboflavin	Niacin	Folate	Phytos (Y or N)
	TOTALS:																			

Recommended Daily Allowances based on 2000 calories per day:

Total fat: 65 g
Saturated fat: 20 g
Trans fat: 0
Cholesterol: 300 mg
Total carbs: 300 g
Fiber: 25 g
Protein: 50 g

Vitamin A: 3000 IU
Vitamin C: 90 mg
Vitamin D: 600 IU
Vitamin E: 15 mg
Calcium: 1300 mg
Iron: 18 mg
Magnesium: 420 mg
Sodium: 2,300 mg (1 teaspoon)

Calorie recommedations based on age, gender and activity level.
(This would be for average height. If you are a lot taller or shorter than average, you need to allow for a few more or a few less calories.)

AGE	sedentary	moderate	active
6-7	1400	1600	1800
8-9	1600	1800	2000
10	1600	1800	2200
11	1800	2000	2200
12	1800	2200	2400
13	2000	2200	2600
14	2000	2200	2600
15	2200	2600	3000
16-18	2400	2800	3200
19-20	2600	2800	3000
21-25	2400	2800	3000
26-35	2400	2600	3000
36-40	2400	2600	2800
41-45	2200	2600	2800
46-50	2200	2400	2800
51-60	2200	2400	2600
61-65	2000	2400	2600
65-70	2000	2200	2600
70+	2000	2200	2400

AGE	sedentary	moderate	active
6-7	1200	1400	1600
8-9	1400	1600	1800
10	1400	1800	2000
11	1600	1800	2000
12	1600	2000	2200
13	1600	2000	2200
14	1800	2000	2400
15	1800	2000	2400
16-18	1800	2000	2400
19-20	2000	2200	2400
21-25	2000	2200	2400
26-35	1800	2000	2400
36-40	1800	2000	2200
41-45	1800	2000	2200
46-50	1800	2000	2200
51-60	1600	1800	2200
61-65	1600	1800	2000
65-70	1600	1800	2000
70+	1600	1800	2000

Definitions of activity levels:
Sedentary= less than 30 minutes a day of moderate physical activity in addition to daily activities
Moderate= at least 30 minutes a day of moderate physical activity
Active= 60 or more minutes a day of moderate physical activity or 30-40 minutes strenuous activity

www.ingramcontent.com/pod-product-compliance
Lightning Source LLC
Chambersburg PA
CBHW081351230426

43667CB00017B/2789